Lecture Notes in Physics

Edited by H. Araki, Kyoto, J. Ehlers, München, K. Hepp, Zürich
R. Kippenhahn, München, D. Ruelle, Bures-sur-Yvette
H. A. Weidenmüller, Heidelberg, J. Wess, Karlsruhe and J. Zittartz, Köln
Managing Editor: W. Beiglböck

335

W0042333

A. Lakhtakia
V. K. Varadan
V. V. Varadan

Time-Harmonic Electromagnetic Fields in Chiral Media

Springer-Verlag
Berlin Heidelberg GmbH

Authors

A. Lakhtakia
V. K. Varadan
V. V. Varadan
Department of Engineering Science & Mechanics
and Research Center for the Engineering of
Electronic and Acoustic Materials
227 Hammond Building, Pennsylvania State University
University Park, PA 16802, USA

ISBN 978-3-662-13770-3   ISBN 978-3-540-46195-1 (eBook)
DOI 10.1007/978-3-540-46195-1

This work is subject to copyright. All rights are reserved, whether the whole or part of the material is concerned, specifically the rights of translation, reprinting, re-use of illustrations, recitation, broadcasting, reproduction on microfilms or in other ways, and storage in data banks. Duplication of this publication or parts thereof is only permitted under the provisions of the German Copyright Law of September 9, 1965, in its version of June 24, 1985, and a copyright fee must always be paid. Violations fall under the prosecution act of the German Copyright Law.

© Springer-Verlag Berlin Heidelberg 1989
Originally published by Springer-Verlag Berlin Heidelberg New York in 1989
Softcover reprint of the hardcover 1st edition 1989

2158/3140-543210 – Printed on acid-free paper

Frontispiece: Maori drawing of a spiral facial tattoo [After R. Huber, *Treasury of Fantastic and Mythological Treasures*, New York: Dover (1981)]

PREFACE

With ongoing progress in the construction of artificial composites, chiral materials will assume great importance. Such materials occur in nature as optically active molecules which display circular birefringence at optical frequencies. Since chirality is a geometric concept, it is conceivable that artificially chiral composites can be made to display circular birefringence even at lower, perhaps microwave, frequencies.

These lecture notes have been prepared to initiate practitioners of electromagnetic theory into the arcana of chiral media. The notes are intended to cover twenty class periods, and have been accordingly arranged. Though Sections 1 through 19 focus exclusively on electromagnetic fields, an extension of these concepts to acoustic waves is given in Section 20 in order to emphasize that it is geometry which gives rise to chirality. An $\exp[-i\omega t]$ harmonic time dependence is implicit throughout this work; only in Section 6 will time-dependent fields be considered. Familiarity with basic electromagnetic theory at the graduate level is assumed, and some experience with dyadic algebra will be useful but not necessary. As such, these notes are suitable for advanced graduate students as well as for researchers with background in electromagnetic theory. We trust that these notes will inspire a considerable number of researchers to be as fascinated by chiral media as we are.

A note on the notation used here: Generally, new notation has been explained upon introduction. An $\exp[-i\omega t]$ harmonic time dependence is implicit throughout, the sole exception being in Section 6. Boldface letters have been used to denote vectors, while German letters denote dyadics. The symbols $\{E, H, D, B\}$ have been used consistently to denote complex-valued field amplitudes; in the first part of Section 6, however, these symbols denote complex instantaneous fields. The symbol σ has been used as a wavenumber in defining wavefunctions, such as $M_n^{(1)}(\sigma r)$ and $N_n^{(1)}(\sigma r)$ of Section 10, as well as the scalar Green's function $g(\sigma:R) = \exp[i\sigma R]/R$. The wavenumbers in an isotropic chiral medium are denoted by γ_1 for the left-circularly polarized fields, and by γ_2 for the right-circularly polarized fields; it is to be emphasized here that $k = \omega(\varepsilon\mu)^{1/2}$ is not a wavenumber in a chiral medium. Throughout this work, it has been assumed that the chiral media are homogeneous as well as at rest. Special notation has been used for acoustically chiral solids in Section 20, wherein it has been explained as necessary.

We wish to record our gratitude to our colleague Professor Richard P. McNitt, who went through the draft manuscript very carefully. Any residual errors are due to us, and no responsibility resides with him!

Comments on the original manuscript provided by Professor Jürgen Ehlers, an editor of the Lecture Notes in Physics Series, resulted in considerable improvements and are greatly appreciated.

This work was prepared under the aegis of the Research Center for the Engineering of Electronic and Acoustic Materials at the Pennsylvania State University, and we thank the industrial sponsors of the Center for their gracious support.

We also acknowledge the patience of our family members, and dedicate this work to a glorious future for all the children of the world.

University Park,   *Akhlesh Lakhtakia*
Pennsylvania.   *Vijay K. Varadan*
March 1989   *Vasundara V. Varadan*

TABLE OF CONTENTS

1. INTRODUCTION

At the beginning of the last century, the phenomenon of optical activity in certain biological and mineral substances was discovered independently by Arago [1811] and Biot [1812, 1815, 1817]. Condensed matter theories were still in embryonic stages at that time. As such, it is remarkable that Pasteur [1848, 1850] interpreted Biot's observations by imagining that the arrangement of atoms within an optically active material is asymmetric in having a non-superposable mirror image; i.e., this arrangement is handed or chiral. Another twenty odd years had to elapse for Pasteur to be vindicated by the simultaneous proposals of the tetravalent carbon atom by van't Hoff [1874] and Le Bel [1874].

Not only did these discoveries require the identification of that property of the optically active substances which causes rotation of polarized light, but they also forced the then scientific establishment to enquire more vigorously into the nature of light. In 1819, Fresnel [1866] had been able to show that the effects of polarized light could be explained by the hypothesis that light was a transverse wave. With advances in electromagnetic field theory which proved that light was indeed an electromagnetic wave, the field of stereochemistry expanded radically and has provided insights into the structure of molecules and, indeed, into the nature of life. For an interesting historical development of this area in the 19th century, a recent review by Applequist [1987] is recommended.

A chiral medium is characterized by either a left-handedness or a right-handedness in its microstructure. As a result, in a chiral medium left- and right- circularly polarized (LCP and RCP) fields propagate with different phase velocities: the field with the latter polarization traveling through a right-handed medium faster than the left-circularly polarized field, and *vice versa*. Optical activity, which is exhibited by a multitude of organic molecules at optical frequencies, is a manifestation of the native chirality of these molecules [Eyring et al 1944; Charney 1979; Barron 1982]. The twin phenomena of circular dichroism (CD) and optical rotatory dispersion (ORD) are observed when a linearly-polarized plane wave travels through a suspension of chiral molecules. ORD refers to the rotation of the plane of polarization of the transmitted planewave with respect to that of the incident planewave, while CD is the differential absorption of the left- and right- circularly polarized planewaves inside the chiral medium. Since these measures of optical activity are material-specific, in the last century and a half CD and ORD measurements have been used extensively by physical chemists to characterize molecular structure [Gray 1916, Lowry 1935].

Gyrotropic studies possess a wealth of information, which has caused Barron [1982] to exclaim that "optical activity provides a peephole into the fabric of universe." Atoms are now considered to be chiral due to the parity-violating weak neutral current interaction between the nucleus and the electrons [Hegstrom et al 1988]; the resulting small optical activity for atoms has quantitative agreement between theory and experiment [Crowe et al 1980]. Recently, the observation of optical activity in atomic vapors has tested the standard model of electroweak unification [Bouchiat & Pottier 1986]. Technological advances in the last two decades have also made possible the detection of chiral asymmetry in Raman scattering [Barron et al 1973; Hug et al 1975]. But molecular light-scattering processes such as Raman scattering, however, involve weak light fluxes and entail phton counting techniques. On the other hand, specular scattering from an achiral/chiral interface entails coherent molecular scattering which results in a strongly scattered wave. Since the ability to observe chiral asymmetries in specular reflection would have strong possibilities as a spectroscopic method for investigating absorbing gyrotropic media, Silverman and his colleagues [Silverman & Black 1987; Silverman et al 1988] have devised experimental techniques correct to 1 part in 10^6 and are currently refining their process further.

Though the quantum-mechanical aspects of chirality have been given much thought [e.g., Barron 1982], a systematic study of the classical electromagnetic field theory in isotropic chiral media has been lacking. In fact, traditionally well-known texts on electromagnetic fields generally ignore chirality. With modern advances in polymer science, however, there is reason to believe that chiral dielectrics, active at the mm-wave frequencies, may become feasible [Corley & Vogl 1980; Harris & Vogl 1981; Vacatello & Flory 1984; Heppke et al 1986]. If polymers with chiral conformations active at sub-optical frequencies become easily realizable, then it will be necessary to examine all aspects of the pertinent electromagnetic field-theoretic relationships. Thus, the feasibility of constructing chiral polymers active at sub-optical frequencies is a prime mover for the present work.

2

In addition, because the specific attribute of chirality is geometry, understanding gleaned from studying molecular structure can be translated into the design and manufacture of artificially chiral media which would exhibit CD and ORD, albeit at the lower frequencies. In this connection, the narrow frequency range of optical activity is noteworthy: at sub-optical frequencies the molecular dimensions are too small compared with the ambient wavelength and optical activity is not observed, while at super-optical frequencies classical electromagnetic concepts must yield to quantum theory. Only in the intermediate range does molecular chirality not vanish when the transition from the microscopic to the macroscopic length scales is effected in electromagnetic theory. These remarks imply that artificial media can be constructed by embedding microminiature chiral objects in an achiral host medium. At lower frequencies the handedness of the microstructure may not be appreciable, and at higher frequencies the microstructure would be so large as to negate the concept of the effective (macroscopic) continuum. But in some intermediate frequency regime, the microstructure size would be a sizable fraction (say 2-5%) of the ambient wavelength in the host medium; consequently, the composite medium would appear to be effectively chiral. The feasibility of this idea renders a study of electromagnetic wave propagation in chiral media very attractive.

Although natural optically active media are isotropic, research on the electromagnetic theory of chiral media may have some impact on vision research as well. The eye contains two types of active structures to distinguish between photopic (high threshold) and scotopic (low threshold) processes: respectively, these are the rods and the cones. Several different types of media in the eye have been known to be anisotropic [deVries et al 1953]; in particular, the retina fibers are thought to be uniaxial [Bour & Lopes Cardozo 1981] while a recent model of the cornea has posited it to be similar to biaxial crystals [van Blokland & Verhelst 1987]. Structural anisotropy has been used by deVries et al [1953] to explain the difference in the sensitivity of eyes to left- and right- circularly polarized light. However, it has been shown that when a rod is illuminated end-on with linearly polarized light, dichroism is not observed because the chromophore dipole of rhodopsin rotates; both vertically and horizontally polarized light components are absorbed equally [Shichi 1983]. Also, biological media are made up of

3

macromolecules which are generally handed. Thus, the retinylidene chromophore of rhodopsin has two distinct circular dichroism bands at 335 and 487 nm [Shaw 1972; Shichi 1971]. In fact, the circular dichroism measurements of rhodopsin suggest that the visual pigments contain a high α-helical content [Shichi 1983]. Furthermore, the primary stroma of the cornea is composed of collagen, and the individual collagen molecule has chirality [Piez 1984]. The triple helix formed from three collagen α-chains is right-handed; hence, fibrils and most tissues, layers and bundles must be handed. However, the issue is probably even more complex, with the handedness possibly alternating from left to right in consecutive layers [Trelstad 1982]. Thus, not only should any treatment of the physics of the eye include anisotropy due to the microstructure -- rods, cones, ganglions, etc., -- but it should also consider the nanostructure, e.g., the helicities of the component molecules whose dimensions may be significant fractions of the optical wavelengths.

As will be shown in the sections to follow, optical activity can be explained by the direct substitution of new constitutive equations, e.g., $\mathbf{D} = \varepsilon\mathbf{E} + \beta\varepsilon\nabla\times\mathbf{E}$ and $\mathbf{B} = \mu\mathbf{H} + \beta\mu\nabla\times\mathbf{H}$, into Maxwell's equations. Here, ε and μ are the usual permittivity and permeability, respectively, while β is a chirality parameter; all three are macroscopic quantities, and β results directly from any chirality in the nanostructure of the medium. As such, the chiral constitutive equations are applicable to any region of the electromagnetic spectrum dealing with non-ionizing radiation. Till quite recently in optics, only intensity measurements were possible; i.e., only the magnitude, and not the phase, of any field could be measured. Thus, the literature on optical activity is generally concerned only with the differences in the intensity of the scattered light when a chiral volume is irradiated by either a LCP or a RCP plane wave. This means that only measurements of $(n_L - n_R)$ are available, where n_L and n_R are the refractive indices for the propagation of LCP and RCP waves in the chiral medium, respectively. Although each of these refractive indices can be related quite readily to ε, μ and β, knowledge of $(n_L - n_R)$ is not sufficient to infer the values of the constitutive parameters. It is quite surprising that optical activity has been known for over a century and a half, but still there are virtually no measured values of the macroscopic material constant that is directly responsible for this phenomenon. The only reference to measured values of β is a paper by Urry and Krivacic [1970], also referred to by Bohren [1975].

4

Were one to consider this phenomenon at the much lower frequency range of 0.5-100 GHz, one would be more concerned about ε, μ and β than about n_L and n_R. With the advent of vector network analyzers, it has now become possible to make very accurate magnitude and phase measurements, but the generation of circularly polarized waves is not always practically possible at these frequencies.

The first attempts to study the effect of relatively large length scale chiral microstructures were by Lindman [1920, 1922], who measured the rotation of the plane of polarization of a linearly-polarized planewave after passage through an ensemble of copper helices of one handedness embedded in a light-weight foam. Analogous measurements were reported much later also by Tinoco and his colleagues [1957, 1960]. These experiments were of a qualitative nature, but are sufficient to warrant a change of name from *optical activity* to *electromagnetic activity*.

Perhaps, the foremost requirement for the measurement of β at microwave frequencies is a departure from the conventional waveguide setup to a free space setup. In a free space setup, planar samples containing chiral microstructures are irradiated by plane waves. The use of a vector network analyzer, coupled with suitably defined canonical experiments, may then facilitate measurements of β. Recent research at Penn State [Guire et al 1988] suggests that reports of measured values of β for artificially chiral composites in the 8-40 GHz range may shortly become available.

2. SCATTERING BY HELICAL ENSEMBLES

At least as important as the actual measurement of the chirality parameter of a given material is the exact realtionship between this parameter and the geometry of the chiral microstructure. In this respect, the *missing link* is not too different from the lack of equations that relate the dielectric constant of a composite material to its microstructure [e.g., Ishimaru 1978]. It is well-known that structure-property relations for ε and μ are easily obtained only if one is willing to describe the microstructure by an electric or a magnetic dipole approximation. In the case of a chiral composite the need for such relations is all the more severe, since chiral microstructures must be artificially introduced for *electromagnetic activity* at microwave frequencies; there are yet no known natural substances which are electromagnetically active at such frequencies. Thus, the material scientist must know how to optimize β of a given chiral composite for a specific application by tailoring the microstructure. Since *tailored materials* are currently considered as requisites for several novel applications, a systematic study of the relationship of the chiral parameter to the microstructure is very desirable.

Helices are perhaps the most common chiral objects found in nature. They are found as seashells [Illert 1987] and as levo- and dextro- molecules [Brand & Fisher 1987]. Very importantly, they form the basis for the α-helix model of many organic molecules which has been much studied by chemists [Cochran et al 1952; Setlow & Pollard 1964]. Indeed, it appears that the helix as well as its two-dimensional counterpart, the spiral, may be considered as the canonical chiral structures. All other manifestations of chirality, -- such as Möbius strips, golf clubs, human hands, etc., -- stem from the helix. Thus, the scattering response of a simple helix assumes a fundamental significance. Additionally, it is important therefore to validate the observation that electromagnetic waves do distinguish between chiral scatterers and their mirror images.

By conducting microwave experiments in the 12-34 cm band, Lindman [1920, 1922] found that the wavelength dependence of the ORD of optically active molecules had the same form as that of 9 cm copper helices. Although this claim was disputed by Winkler [1956], Tinoco and his colleagues [1957, 1960] soon demonstrated the truth of Lindman's results. With advances in digital computers, simulation of the scattering by helices has become a very prominent area of research in the physical chemistry community, as is testified by the large number of workers in the field [e.g., Bustamente et al 1982; Belmont et al 1985; Keller et al 1985; Patterson et al 1986]. Emphasis in these studies has been laid on the calculation of the circular intensity differential scattering (CIDS) signatures of helices, CIDS being a normalized difference between the scattered intensity patterns for incident LCP and RCP planewaves averaged over all possible orientations of the helix at a given frequency.

Thus, the significance of chiral media cannot be denied. But chirality does not have to have molecular origins, since it is nothing but geometric. Therefore, as pointed out earlier, effectively chiral media can be constructed by embedding chiral microstructures in non-chiral host media. Such microstructures may be macromolecular polymers with helical conformations [Vacatello & Flory 1984], or even microminiature helical springs *a la* Lindman. What is of interest for electromagnetic (EM) use is the optimization of the helix's chirality: the parameters governing the helix geometry must be examined for their effect on an incident EM wave. Varadan et al [1988] have modeled helical polymers by an arrangement of spherical beads (in themselves, large molecules for the present purposes) suspended on a helical strand which is indistinguishable from the surrounding free space. Let the spheres be sufficiently small, compared to the free space wavelength, that they can be modeled as point electric dipoles.

In a cartesian co-ordinate system, the helix on which the tiny spheres are located is given by the radius vector

$$\mathbf{r}(\xi) = a[\mathbf{e}_x \cos\xi + \mathbf{e}_y h \sin\xi] + \mathbf{e}_z P(\xi/2\pi); \quad \xi \in \{-\infty, \infty\}, \quad (2\text{-}1)$$

where a is the radius and P is the pitch of the helix; the handedness parameter h =+1 if the helix curls up in the +z direction according to the right-handed rule, and h = -1, if otherwise; and $\mathbf{e}_x$, $\mathbf{e}_y$, and $\mathbf{e}_z$ are the unit orthogonal vectors. Let the helix be finite in extent, having 2N+1 complete rotations, N being a positive integer or zero. On each of the 2N+1 rings of this finite helix, there are 2M+1 spheres arranged over equal–$\Delta\xi$ segments, M ≥ 1 being integral. Each of the spheres in this arrangement has a radius b which is small enough that no two of them ever touch.

The spheres possess a dielectric constant ε, and if a field $\boldsymbol{E_m}$ illuminates the mth sphere, an electric dipole moment [Jackson 1975]

$$\boldsymbol{p_m} = \alpha \, \boldsymbol{E_m} = 4\pi\varepsilon_o b^3 \, (\varepsilon - \varepsilon_o)(\varepsilon + 2\varepsilon_o)^{-1} \, \boldsymbol{E_m} \qquad (2\text{-}2)$$

is induced on it, where α is the isotropic electric polarizability of the spheres, ε_o is the permitivitty of free space, and μ_o is the free space permeability.

The EM field incident on the helical ensemble, $\boldsymbol{E}_{inc}$, can be any arbitrary field so long as its source is not located anywhere inside or on the minimum sphere circumscribing the helix. But the field $\boldsymbol{E_m}$ actually incident on the mth sphere is not $\boldsymbol{E}_{inc}(\boldsymbol{r_m})$ alone; it also consists of the fields re-radiated by all of the other spheres as well. With this reasoning, the system of 3Q simultaneous equations [Patterson et al 1986],

$$\boldsymbol{E_m} = \boldsymbol{E}_{inc}(\boldsymbol{r_m}) + \alpha k_o^2 (4\pi\varepsilon_o)^{-1} \sum_{n,n\neq m} \left[R_{mn}^{-1} \exp[ik_o R_{mn}] \right] \cdot$$
$$\cdot \{ g_{mn} \boldsymbol{n}_{mn}(\boldsymbol{n}_{mn} \cdot \boldsymbol{E_n}) - (1/3)[g_{mn} - 2] \, \boldsymbol{E_n} \}], \qquad (2\text{-}3)$$

must be solved, in order to obtain the various exciting fields $\boldsymbol{E_m}$. Here, $R_{mn} = |\boldsymbol{r_m} - \boldsymbol{r_n}|$, $\boldsymbol{n}_{mn} = (\boldsymbol{r_m} - \boldsymbol{r_n})/R_{mn}$, $g_{mn} = 3(k_o R_{mn})^{-2} - 3i(k_o R_{mn})^{-1} - 1$, while the free space wavenumber $k_o = \omega\sqrt{(\varepsilon_o\mu_o)}$ for the time-harmonic excitation $\exp[-i\omega t]$. Once the solution of (2-3) has been obtained, the total scattered field for $k_o r \rightarrow \infty$ can be computed from the Fraunhoffer-type relation

$$4\pi\varepsilon_o \boldsymbol{E}_{sc}(\boldsymbol{r}) = \alpha k_o^2 r^{-1} \exp[ik_o r]$$
$$\cdot \sum_m \{ \exp[-ik_o \boldsymbol{r_m} \cdot \boldsymbol{r}/r][\boldsymbol{E_m} - \boldsymbol{r}(\boldsymbol{r} \cdot \boldsymbol{E_m})/r^2] \}. \qquad (2\text{-}4)$$

Multiple scattering between the spheres would ensure that $\boldsymbol{E_m} \neq \boldsymbol{E}_{inc}(\boldsymbol{r_m})$; in other words, the field actually incident on the m-th sphere is different from the incident field at that location. It is to be expected that the exciting fields should reflect the chirality of the helical ensemble, because the ensemble of unit vectors $\{\boldsymbol{n}_{mn}\}$ used in (2-4) is handed, even though the scalar distances $\{R_{mn}\}$ do not depend on h. This, indeed turned out to be the case. From the computations a very interesting conclusion can be drawn: provided $\boldsymbol{k}_{inc} \parallel \boldsymbol{e_x}$ or $\boldsymbol{k}_{inc} \parallel \boldsymbol{e_z}$, then (i) only

$e_y \cdot \boldsymbol{E}_m$ changes sign with h when $e_y \cdot \boldsymbol{E}_{inc} = 0$; and (ii) only $e_y \times \boldsymbol{E}_m$ changes sign with h when $e_y \times \boldsymbol{E}_{inc} = 0$. Symmetries of similar natures are also present when $k_{inc} \parallel e_y$, but they are not so easily describable.

Since the source of the scattered field $\boldsymbol{E}_{sc}$ is not concentrated at any given location, it can be conveniently expressed as being due to a set of multipoles concentrated at the origin. The two lowest order multipoles are the electric dipole, p_{eqvt}, and the magnetic dipole, m_{eqvt}. The equivalent electric dipole of the helical ensemble can be computed by taking the projection of $\boldsymbol{E}_{sc}$ on $\boldsymbol{E}_p$ in the limit $k_o r \rightarrow \infty$ as

$$\int_0^\pi d\theta \, \sin\theta \int_0^{2\pi} d\varphi \, \boldsymbol{E}_p(r) \cdot \boldsymbol{E}_p{}^*(r) = $$
$$\int_0^\pi d\theta \, \sin\theta \int_0^{2\pi} d\varphi \, \boldsymbol{E}_{sc}(r) \cdot \boldsymbol{E}_p{}^*(r), \qquad (2\text{-}5a)$$

in which $\boldsymbol{E}_p$ is the electric field radiated by the equivalent electric dipole in the far zone,

$$4\pi\varepsilon_o \boldsymbol{E}_p(r) = -k_o{}^2 \, r^{-3} \, \exp[ik_o r] \, r \times \{r \times p_{eqvt}\}. \qquad (2\text{-}5b)$$

Similarly, the magnetic dipole moment can be calculated by taking the projection of $\boldsymbol{E}_{sc}$ on $\boldsymbol{E}_m$ in the limit $k_o r \rightarrow \infty$ as

$$\int_0^\pi d\theta \, \sin\theta \int_0^{2\pi} d\varphi \, \boldsymbol{E}_m(r) \cdot \boldsymbol{E}_m{}^*(r) = $$
$$\int_0^\pi d\theta \, \sin\theta \int_0^{2\pi} d\varphi \, \boldsymbol{E}_{sc}(r) \cdot \boldsymbol{E}_m{}^*(r), \qquad (2\text{-}6a)$$

where $\boldsymbol{E}_m$ is the electric field radiated by the equivalent magnetic dipole in the far zone and is given by

$$4\pi\varepsilon_o \boldsymbol{E}_m(r) = -k_o{}^2 \, [\mu_o\varepsilon_o]^{1/2} \, r^{-2} \, \exp[ik_o r] \, \{r \times m_{eqvt}\}. \qquad (2\text{-}6b)$$

The net result of these exertions can be stated as a sum of matrix-vector products as

$$\begin{bmatrix} e_x \cdot P_{eqvt} \\ e_y \cdot P_{eqvt} \\ e_z \cdot P_{eqvt} \end{bmatrix} = 3\alpha \sum_m \begin{bmatrix} I_{11}(r_m) & I_{12}(r_m) & I_{13}(r_m) \\ I_{12}(r_m) & I_{22}(r_m) & I_{23}(r_m) \\ I_{13}(r_m) & I_{23}(r_m) & I_{33}(r_m) \end{bmatrix} \begin{bmatrix} e_x \cdot E_m \\ e_y \cdot E_m \\ e_z \cdot E_m \end{bmatrix} , \qquad (2\text{-}7)$$

and

$$\begin{bmatrix} e_x \cdot m_{eqvt} \\ e_y \cdot m_{eqvt} \\ e_z \cdot m_{eqvt} \end{bmatrix} = 3\alpha \sum_m \begin{bmatrix} 0 & J_{12}(r_m) & -J_{13}(r_m) \\ -J_{12}(r_m) & 0 & J_{23}(r_m) \\ J_{13}(r_m) & -J_{23}(r_m) & 0 \end{bmatrix} \begin{bmatrix} e_x \cdot E_m \\ e_y \cdot E_m \\ e_z \cdot E_m \end{bmatrix} . \qquad (2\text{-}8)$$

The various entries in the foregoing matrices can be analytically calculated out as

$$I_{11}(r_m) = 2\, j_1(k_o r_m)/k_o r_m - (1 - \sin^2\theta_m \cos^2\varphi_m)\, j_2(k_o r_m), \qquad (2\text{-}9a)$$

$$I_{22}(r_m) = 2\, j_1(k_o r_m)/k_o r_m - (1 - \sin^2\theta_m \sin^2\varphi_m)\, j_2(k_o r_m), \qquad (2\text{-}9b)$$

$$I_{33}(r_m) = 2\, j_1(k_o r_m)/k_o r_m - \sin^2\theta_m\, j_2(k_o r_m), \qquad (2\text{-}9c)$$

$$I_{12}(r_m) = \sin\varphi_m \cos\varphi_m \sin^2\varphi_m\, j_2(k_o r_m), \qquad (2\text{-}10a)$$

$$I_{13}(r_m) = \cos\varphi_m \sin\theta_m \cos\theta_m\, j_2(k_o r_m), \qquad (2\text{-}10b)$$

$$I_{23}(r_m) = \sin\varphi_m \sin\theta_m \cos\theta_m\, j_2(k_o r_m), \qquad (2\text{-}10c)$$

$$J_{12}(r_m) = i\, [\mu_o \varepsilon_o]^{-1/2} \cos\theta_m\, j_1(k_o r_m), \qquad (2\text{-}11a)$$

$$J_{13}(r_m) = i\, [\mu_o \varepsilon_o]^{-1/2} \sin\varphi_m\, j_1(k_o r_m), \qquad (2\text{-}11b)$$

$$J_{23}(r_m) = i\, [\mu_o \varepsilon_o]^{-1/2} \cos\varphi_m\, j_1(k_o r_m), \qquad (2\text{-}11c)$$

in which $r_m \equiv \{r_m, \theta_m, \varphi_m\}$ is the position vector of the m-th sphere of the helical ensemble in the spherical co-ordinate system; and $j_1(\cdot)$, etc. are the spherical Bessel functions. It is to be noted here that the handedness of the helical ensemble has already been taken into account during the calculation of the exciting fields E_m.

Since the helical ensemble is much smaller than the ambient wavelength, a low-frequency approximation of the matrices in (2-7) and (2-8) can be used for further analysis. Suppose the pitch P were to be zero: then the diagonal terms of the matrix in (2-7) are of the lowest order in $k_o a$, while the quantities defined in (2-10) and (2-11) are of higher orders. Consequently, the equivalent magnetic dipole

10

moment is small enough to be ignored. Likewise, let the helix be straightened out into a straight line such that the radius a becomes zero; remembering that the overall helix dimensions are small leads to the same conclusion as well. Both of these conclusions are physically valid: a small ring of point electric dipoles will appear to be a point electric dipole itself, and a straight-line ensemble will look like a short electric dipole, in the far zone. It appears, therefore, that optimally chiral helical strands must have maximal values of $\mathbf{m}_{eqvt}$.

Several studies were carried out [Varadan et al 1988] for different planewave incidence conditions and the results of these investigations are summarized now. From studying the magnitude of $\mathbf{m}_{eqvt}$ as a function of the ratio a/P for several different incident plane waves, it was observed that when a/P ~ 0.24, optimal chirality is achieved because $\mathbf{m}_{eqvt}$ is optimally high. The computed *optimal* value of the ratio a/P is, interestingly enough, not far from values reported for various helical polymers [Miyazawa 1961; Setlow & Pollard 1964]. It is important to realize that chirality will not be observed if a/P = 0 or P/a = 0, because in these two respective limits the ensemble of scatterers is either linear or circular, and thus, devoid of handedness. Because polymers in the helical configuration deviate only slightly from their linear counterparts, on the other hand, the ratio a/P cannot be very large either [Elert & White 1987]. Thus, the optimal value of a/P = 0.24 is quite satisfactory.

Furthermore, it is obvious that by increasing α, $\mathbf{m}_{eqvt}$ can also be enhanced due to the enhancement in the interactions between the dipoles: but the spheres are small and α is only weakly dependent on $\varepsilon/\varepsilon_o$. Even so, α should be as large as possible. Next, it was also found that increasing the number of spheres per ring (i.e., increasing M) is also helpful in increasing chirality by enhancing the number of dipole-dipole interaction paths. From the numerical studies it turned out that short helices appeared to be preferable to large ones, i.e., N should be as small as possible; in other words, a larger number density of fewer-turn helices may be preferable to a smaller number density of many-turn ones.

The foregoing observations should stand in good stead for designers of chiral materials: either as artificial media having chiral nanostructure, or as polymers composed of chiral macromolecules. Artificial media are constructed by embedding small particles of one material into a matrix medium: in some frequency range, the composite appears as an *effective* medium for coherent wave propagation. Theoretical analyses, generally lumped as *effective medium theories* [Ishimaru 1978], do not address the handedness of the particle shape, being largely concerned with the volumetric concentrations of the two phases. The only procedure the authors know of, which involves the particle shape explicitly is based on the T-matrix method [Varadan & Varadan 1980]. It is anticipated that with further interest in chiral composites, the handedness of the embedded particles will be used as an additional control parameter for realizing composites with specific properties.

3. CONSTITUTIVE EQUATIONS

Optically active media (mostly organic materials) are naturally chiral at optical frequencies; interesting examples include the famous Watson-Crick double-helix representation of the DNA molecule. But the underlying principle behind chirality is the mirror-asymmetry of the constituent microstructure. Therefore, artificially chiral media can be constructed by embedding chiral microstructures in non-chiral host media, such composites being effectively chiral even at the sub-optical, and the high mm-wave, frequencies. The microstructure size should be large enough (compared to the wavelength in the matrix medium) so that the spatial variation of the EM field can sense its handedness; at the same time, it should be small enough so that, at least in some frequency range, the composite should appear to be effectively chiral.

In order to understand the characteristics of chiral media, whether artificially so or naturally, one has to begin with the notions of the intrinsic polarization and the intrinsic magnetization vectors. Since free space is vacuous, it does not contain any matter. The constitutive relations for the electromagnetic fields in free space can be adequately expressed in the form $\mathbf{D} = \varepsilon_o\mathbf{E}$ and $\mathbf{B} = \mu_o\mathbf{H}$. Fields in material media induce bound charges. Collectively, the constitutive equations in material media have the form

$$\mathbf{D} = \varepsilon_o\mathbf{E} + \mathbf{P}, \tag{3-1a}$$
$$\mathbf{B} = \mu_o\mathbf{H} + \mathbf{M}, \tag{3-1b}$$

in which $\mathbf{P}$ and $\mathbf{M}$, respectively, are the polarization and the magnetization.

The polarization $\mathbf{P}$ is interpreted as the average electric dipole moment per unit volume, while $\mathbf{M}$ is the average magnetic dipole moment per unit volume. It is important to realize that $\mathbf{E}$, $\mathbf{P}$ and $\mathbf{D}$ are *polar* vectors, while $\mathbf{H}$, $\mathbf{M}$ and $\mathbf{B}$ are *axial* vectors. The components of a polar vector change sign under a spatial inversion of the coordinate system, whereas those of an axial vector do not. This turns out to be of special significance in determining the natures of the constitutive parameters.

In non-magnetic dielectric media, a prominent contribution to $\mathbf{P}$ comes from the charge separation in the atom, and is designated as *electronic* polarization. This is because an applied electric field causes displacement of the electrons with respect to the atomic nucleus. *Atomic* or *ionic* polarization may be caused by the displacement of atoms or ions in a molecule by electric fields. Some media may also have permanent electric dipole moments because the centers of the negative and the positive charges are not co-incident. Without an applied electric field, these permanent dipoles are randomly aligned, and their effect is nulled on the average; however, when an electric field is applied, these dipole moments line up to give rise to *orientational* polarization. In linear media, these various contributions can be summed up as $\mathbf{P} = \chi_e \varepsilon_o \mathbf{E}$, so that

$$\mathbf{D} = \varepsilon_o[1 + \chi_e]\mathbf{E} = \varepsilon\mathbf{E}, \qquad\qquad (3\text{-}2)$$

in which χ_e is the electric susceptibility; and ε, the (real) dielectric constant is a true scalar. Under the influence of the externally applied field, the polar molecules rotate towards an equilibrium distribution. If the polar molecules are massive, or if the frequency is very high, the rotatory motion lags behind the applied field and equilibrium is never attained. Thus, the polarization is no longer in sync with the applied field, giving rise to a conduction current density. This results in thermal dissipation of energy, which phenomenon is known as Ohm's law. The ohmic loss in a dielectric medium is quantified in terms of a conductivity, which can be incorporated in the time-harmonic constitutive equations by making ε complex.

There are microscopic currents due to electron spin and the motion of electrons around the nucleus. These microscopic currents act as sources of macroscopic magnetic fields, thereby endowing the electrons and the atoms with magnetic dipole moments; but these currents do not produce macroscopic charge transport. In magnetic media, the effect of the magnetic dipole moments cannot be ignored, giving rise to the magnetization $\mathbf{M}$. Again, for linear media, it can be stated that $\mathbf{M} = \chi_m \mu_o \mathbf{H}$ so that

$$\mathbf{B} = \mu_o[1 + \chi_m]\mathbf{H} = \mu\mathbf{H}, \qquad\qquad (3\text{-}3)$$

χ_m being the magnetic susceptibility, and the true scalar μ being the (real) permeability. Magnetic ohmic losses can be incorporated by making μ complex in the same fashion as the electric ohmic losses.

As stated earlier, chirality was first observed as optical activity, which is the rotation of the plane of polarization in certain linear isotropic media. Phenomenological studies by Drude [1900] indicated that the rotation of the plane of polarization is predicted by Maxwell's equations provided **P** has an additional term proportional to $\nabla \times \mathbf{E}$. These considerations lead Born [1915, 1972] to the proposal that

$$\mathbf{D} = \varepsilon \, [\mathbf{E} + \eta \nabla \times \mathbf{E}], \qquad\qquad\qquad (3\text{-}4a)$$
$$\mathbf{B} = \mu \, \mathbf{H}, \qquad\qquad\qquad (3\text{-}4b)$$

the psudoscalar η being the chirality parameter. A medium following the relations (3-4) is non-reciprocal [Bohren 1975a; Satten 1958]. In addition, after studying the reflection and transmission characteristics of planar achiral-chiral interfaces, Silverman [1986] has shown that the use of (3-4) fails to satisfy the standard boundary conditions and results in physically unacceptable amplitudes around the critical angles of incidence. Born's proposal for isotropic chirality has been modified [Fedorov 1959a,b; Bokut' & Fedorov 1959] to

$$\mathbf{D} = \varepsilon \, [\mathbf{E} + \beta \nabla \times \mathbf{E}], \qquad\qquad\qquad (3\text{-}5a)$$
$$\mathbf{B} = \mu \, [\mathbf{H} + \beta \nabla \times \mathbf{H}], \qquad\qquad\qquad (3\text{-}5b)$$

which are not only symmetric under time-reversality [Satten] and duality transformations [Silverman 1986] but also satisfy the tests set up by Silverman [1985] to distinguish between (3-4) and (3-5); the pseudoscalar β is the measure of chirality here, and it carries the unit of a length. The validity of (3-5) has been affirmed by studies carried on optically active molecules [Bohren 1976] as well as from the examination of light propagation in optically active crystals [Fedorov 1959a,b; Bokut' & Fedorov 1959]. The nonlocal character [Eringen 1984] of (3-5) needs to be noticed, because the polarization **P** (resp. magnetization **M**) depends not only on **E** (resp. **H**) but also on the *circulation* of **E** (resp. **H**). In a not-too-rigorous manner, one may even observe that **P** (resp. **M**) has a component due to the time-rate of change of **H** (resp. **E**), *vide* Faraday and Ampere-Maxwell laws. In the last decade, Bohren has used the Drude-Born-Fedorov constitutive relations (3-5) to compute the scattering responses of chiral spheres [Bohren 1974], spherical shells [Bohren 1975b] and infinitely long, right circular cylinders [Bohren 1978]. The extensive studies carried out by the Penn State group on chiral media using (3-5) form the basis of these notes.

It is to be noted here that Condon's equations [Condon 1937; Charney 1979] for optical activity, *viz.*,

$$\mathbf{D} = \varepsilon_C \mathbf{E} - \chi \partial \mathbf{H}/\partial t, \tag{3-6a}$$

$$\mathbf{B} = \mu_C \mathbf{H} + \chi \partial \mathbf{E}/\partial t, \tag{3-6b}$$

are equivalent to (3-5), provided harmonic time-dependence, $\exp[-i\omega t]$ is assumed [Bohren 1975a]; in that case, the correspondence between (3-5) and (3-6) is given by $\chi = \varepsilon\mu\beta/(1 - \omega^2\varepsilon\mu\beta^2)$, $\varepsilon_C = \varepsilon/(1 - \omega^2\varepsilon\mu\beta^2)$ and $\mu_C = \mu/(1 - \omega^2\varepsilon\mu\beta^2)$. Furthermore, a take-off from Tellegen's formulation for the gyrator [Tellegen 1948] gives rise to yet another set of constitutive equations [Chambers 1956; Unz 1964; Krowne 1984] for isotropic reciprocal chiral media, *viz.*,

$$\mathbf{D} = \varepsilon_T \mathbf{E} + \zeta \mathbf{H}, \tag{3-7a}$$

$$\mathbf{B} = \mu_T \mathbf{H} - \zeta \mathbf{E}. \tag{3-7b}$$

This set is also equivalent to (3-5) for the time-harmonic case, provided the chirality parameter $\zeta = i\omega\varepsilon\mu\beta/(1 - \omega^2\varepsilon\mu\beta^2)$, $\varepsilon_T = \varepsilon/(1 - \omega^2\varepsilon\mu\beta^2)$ and $\mu_T = \mu/(1 - \omega^2\varepsilon\mu\beta^2)$. Both chirality parameters, χ of (3-6) and ζ of (3-7), are pseudoscalars.

A remarkable set of equations has been deduced by Post [1962] who did not consider the medium microstructure at all, but simply required all equations to be generally covariant. His proposal

$$\mathbf{D} = \varepsilon_P \mathbf{E} + i\xi \mathbf{B}, \tag{3-8a}$$

$$\mathbf{B} = \mu_P [\mathbf{H} - i\xi \mathbf{E}] \tag{3-8b}$$

was also obtained [Jaggard et al 1978] phenomenologically by considering a chiral medium composed of a dilute suspension of perfectly conducting, single-turn helices embedded in an otherwise achiral host medium. The validity of (3-8) has

16

also been tested by studies conducted on optically active molecules [Eyring et al 1944; Post 1962]. These constitutive equations have recently been used by Bassiri et al [1986] to derive an *infinite-medium* Green's function and to calculate the radiation field of a short electric dipole embedded in a chiral medium. Eqs. (3-8) are "similar" to those set up by Cheng and Kong [1968a] for bianisotropic media. The correspondence between (3-5) and (3-8) holds with $\varepsilon_P = \varepsilon$, $\mu_P = \mu/(1-\omega^2\varepsilon\mu\beta^2)$ and the pseudoscalar $\xi = \omega\varepsilon\beta$ for the time-harmonic fields.

Quite a comprehensive exposition of the electromagnetic field theory has been done for bianisotropic media [Cheng & Kong 1968a,b; Kong & Cheng 1968a,b; Kong 1972, 1974], whose constitutive equations are in tensor form and were given by Post [1962] as

$$c\mathbf{D} = \mathbb{P}\cdot\mathbf{E} + c\mathbb{L}\cdot\mathbf{B}, \qquad\qquad (3\text{-}9a)$$
$$\mathbf{H} = \mathcal{M}\cdot\mathbf{E} + c\mathbb{Q}\cdot\mathbf{B} \qquad\qquad (3\text{-}9b)$$

in which c is the speed of light *in vacuo*; and $\mathbb{P}$, etc., are 3×3 matrices in a cartesian co-ordinate system. Parenthetically, it should be observed that while $\mathbb{P}$ and $\mathbb{Q}$ are real tensors, $\mathbb{L}$ and $\mathcal{M}$ are only pseudotensors. A considerable part of this work is summarised in a paper by Kong [1972] wherein he has shown that (3-9) can be recast also in a bianisotropic Tellegen form. For time-harmonic fields, reciprocity and symmetry conditions have been examined, and conditions for lossless propagation have been determined. The conditions for the conservation of energy and electromagnetic momentum have also been established. Lastly, an *infinite-medium* Green's dyadic has been formulated, though it is given as an integral in the (three-dimensional) spatial frequency space. The reduction of the various expressions derived in this paper [Kong 1972] for isotropic chiral media with the isotropic Post constitutive equations is certainly possible. Most of this work, however, has been applied to moving [van Bladel 1984] or stationary, isotropic or biaxial or uniaxial media. Magnetoelectric media, as they are called in materials literature, are also modeled by (3-9); and earlier work on plane wave propagation these media has been summarised in a book by O'Dell [1970].

The merits of the various constitutive equations given above are debatable. Distinctions between them can only be possible for arbitrarily time-dependent, and not simply time-harmonic, processes as well as for inhomogenous media. For these cases, however, not much work seems to have been done. In any event, these considerations fall outside the scope of the present work.

The various constitutive equations have been shown to be equivalent to each other for time-harmonic fields: while the equations of Post, Condon and Tellegen preserve the permeability $|B|/|H|$, (3-5) and (3-8) have the same permittivity $|D|/|E|$. As mentioned above, the substantial work of Kong and his colleagues on bianisotropic media can certainly be applied for studying waves in isotropic chiral media, although the resulting isotropic equations will be of the Post type (3-8). However, *the asymmetry associated with optical activity is immediately apparent in (3-5) as opposed to the other constitutive equations:* the curl is not a vector under a reflection of co-ordinate systems [Altman et al 1984]. Hence, in the sequel it was decided to use the Drude-Born-Fedorov constitutive equations (3-5).

4. FIELD EQUATIONS

The development of the constitutive equations has to be followed by their integration into Maxwell's equations. Though chirality can be introduced into an otherwise achiral medium simply by subjecting it to a static magnetic field [Barron 1982], it is necessary to realize that a chiral volume will exhibit its handedness only while interacting with time-varying fields. In addition, it should be noted also that ε, μ and β may be frequency-dependent; unless otherwise stated, a harmonic time-dependence $\exp[-i\omega t]$ has been assumed throughout this work.

Use of the first two monochromatic Maxwell's equations, $\nabla \cdot \mathbf{D} = 0$ and $\nabla \cdot \mathbf{B} = 0$, in conjunction with the constitutive equations (3-5) easily yields the fact that $\mathbf{E}$ and $\mathbf{H}$ are also divergenceless. Furthermore, using the latter two Maxwell's equations, $\nabla \times \mathbf{E} = i\omega \mathbf{B}$ and $\nabla \times \mathbf{H} = -i\omega \mathbf{D}$, along with the constitutive equations (3-5) and some manipulations, yields the *equivalent monochromatic* constitutive relations

$$(1 - k^2\beta^2) \, \mathbf{D} = \varepsilon \, \mathbf{E} + i(\beta/\omega)k^2 \, \mathbf{H}, \tag{4-1a}$$
$$(1 - k^2\beta^2) \, \mathbf{B} = \mu \, \mathbf{H} - i(\beta/\omega)k^2 \, \mathbf{E}, \tag{4-1b}$$

where $k = \omega[\varepsilon\mu]^{1/2}$ is simply a shorthand notation and does not represent any wavenumber inside the chiral medium. It must be mentioned that (4-1) are similar to the isotropic Tellegen equations (3-7). Four other relations dealing with the circulation of the field quantities may also be derived for later use from (3-5) and (4-1):

$$\nabla \times \mathbf{E} = \gamma^2\beta \, \mathbf{E} + i\omega\mu \, (\gamma/k)^2 \, \mathbf{H}, \tag{4-2a}$$
$$\nabla \times \mathbf{H} = \gamma^2\beta \, \mathbf{H} - i\omega\varepsilon \, (\gamma/k)^2 \, \mathbf{E}, \tag{4-2b}$$
$$\nabla \times \mathbf{D} = \gamma^2\beta \, \mathbf{D} + i\omega\varepsilon \, (\gamma/k)^2 \, \mathbf{B}, \tag{4-3a}$$
$$\nabla \times \mathbf{B} = \gamma^2\beta \, \mathbf{B} - i\omega\mu \, (\gamma/k)^2 \, \mathbf{D}. \tag{4-3b}$$

In these equations, the parameter γ is given by

$$\gamma^2 = k^2 \, [1 - k^2\beta^2]^{-1}. \tag{4-4}$$

19

To develop a Helmholtz-like equation for $\mathbf{E}$, the curl of both sides of (4-2a) gives the relation

$$-\nabla\times\nabla\times\mathbf{E} + \gamma^2\beta\,\nabla\times\mathbf{E} + i\omega\mu\,(\gamma/k)^2\,\nabla\times\mathbf{H} = 0. \tag{4-5}$$

Then, Ampere's law, $\nabla\times\mathbf{H} = -i\omega\mathbf{D}$, is substituted into (4-5) and use is made of (3-5b) to obtain the differential equation which *governs* the electric field:

$$-\nabla\times\nabla\times\mathbf{E} + 2\gamma^2\beta\,\nabla\times\mathbf{E} + \gamma^2\,\mathbf{E} = 0. \tag{4-6}$$

By utilizing similar manipulations, it can be shown that (4-6) is satisfied by all of the fields, *viz.*,

$$-\nabla\times\nabla\times\mathbf{U} + 2\gamma^2\beta\,\nabla\times\mathbf{U} + \gamma^2\,\mathbf{U} = 0; \qquad \mathbf{U} = \mathbf{E},\,\mathbf{H},\,\mathbf{D},\,\mathbf{B}. \tag{4-7}$$

It should be mentioned that hereafter $\mathbf{U}$ will denote any of these four fields, or any linear combination thereof. Also to be noted is the fact that when $\beta = 0$, (4-7) reduces to the vector Helmholtz equation, $\nabla^2\mathbf{U} + k^2\mathbf{U} = 0$, for non-chiral media .

For radiation problems, (4-7) is not enough since the source terms are absent. The source-incorporated Maxwell's equations,

$$\nabla\times\mathbf{E} = i\omega\mathbf{B} - \mathbf{K}, \tag{4-8a}$$
$$\nabla\times\mathbf{H} = -i\omega\mathbf{D} + \mathbf{J}, \tag{4-8b}$$

in which $\mathbf{J}$ and $\mathbf{K}$, respectively, are volume electric and magnetic current densities, along with the constitutive equations (3-5), can be manipulated to yield the differential equations:

$$\nabla\times\mathbf{E} - \gamma^2\beta\,\mathbf{E} - i\omega\mu(\gamma/k)^2\mathbf{H} = i\omega\mu\beta(\gamma/k)^2\mathbf{J} - (\gamma/k)^2\mathbf{K}, \tag{4-9a}$$
$$\nabla\times\mathbf{H} - \gamma^2\beta\,\mathbf{H} + i\omega\varepsilon(\gamma/k)^2\mathbf{E} = i\omega\varepsilon\beta(\gamma/k)^2\mathbf{K} + (\gamma/k)^2\mathbf{J}. \tag{4-9b}$$

Equations (4-9) are further manipulated using (3-5) and (4-8) in order to finally obtain the governing differential equations for $\mathbf{E}$ and $\mathbf{H}$, respectively, as

20

$$\nabla \times \nabla \times \mathbf{E} - 2\gamma^2\beta \, \nabla \times \mathbf{E} - \gamma^2 \mathbf{E} = i\omega\mu(\gamma/k)^2[\mathbf{J} + \beta\nabla \times \mathbf{J}] - (\gamma/k)^2\nabla \times \mathbf{K}, \quad (4\text{-}10a)$$

$$\nabla \times \nabla \times \mathbf{H} - 2\gamma^2\beta\nabla \times \mathbf{H} - \gamma^2 \mathbf{H} = i\omega\varepsilon(\gamma/k)^2[\mathbf{K} + \beta\nabla \times \mathbf{K}] + (\gamma/k)^2\nabla \times \mathbf{J}. \quad (4\text{-}10b)$$

Of more than cursory importance are the boundary conditions to be used for bimaterial interfaces involving at least one chiral medium. These have been the subject of an intense debate among the proponents of the various sets of constitutive equations; see, for example, Hornreich & Shtrikman [1968], Bokut' & Serdyukov [1972], and Agranovich & Ginzburg [1973]. The integral forms of the Faraday and the Ampere-Maxwell equations imply that boundary conditions must be prescribed on the tangential components of the $\mathbf{E}$ and the $\mathbf{H}$ fields for time-varying problems, these conditions being both necessary and sufficient. The boundary conditions on the normal components of $\mathbf{D}$ and $\mathbf{B}$, being scalar equations are necessary and sufficient for static problems, but are not sufficient for general time-varying problems. Knowing that chirality does not exist for static problems, it is very satisfying that Maxwell's equations give rise to the sufficient boundary conditions which involve only $\mathbf{E}$ and $\mathbf{H}$ fields.

Yet another issue needs to be addressed here: the status of the magnetic current density $\mathbf{K}$. Maxwell's equations with $\mathbf{K} = 0$ are Abelian and possess a U(1) gauge symmetry. When $\mathbf{K} \neq 0$ is incorporated, Maxwell's equations become non-Abelian and possess the SU(2) symmetry. U(1) fields have less local degrees of freedom than SU(2) fields, but the latter can be transformed into the former through a process known as *symmetry breaking* [O'Raifeartaigh 1975]. As will be shown later in Sec. 15, both electric and magnetic sources are needed simultaneously to produce the allowed eigenstates (left- and right- circular) of polarization in chiral media. In addition, $\mathbf{K}$ turns out to be of considerable use in establishing the duality principle [Jackson 1975], which will be established for chiral media in the following section.

21

5. REACTION, RECIPROCITY AND DUALITY

The equations developed in the previous section will be utilized at a later stage for setting up infinite-medium Green's functions and Huygens's principle for isotropic chiral media. In the interim, however, attention will be focused on the derivation of some intrinsic field-theoretic relationships.

To begin with, consider a chiral medium $(\varepsilon, \mu, \beta)$ in which the electric and the magnetic sources $\{J_a, K_a\}$ give rise to the fields $\{E_a, H_a, B_a, D_a\}$; and another set of sources $\{J_b, K_b\}$ which independently create the fields $\{E_b, H_b, B_b, D_b\}$. The sources can be distributed or localized, but cannot be spread over all space and are generally located at a finite distance from the origin. Then, as per (4-9), provided both sets of sources operate at the same frequency,

$$\nabla \times E_{a,b} = \gamma^2 \beta \, E_{a,b} + i\omega\mu(\gamma/k)^2 H_{a,b} + i\omega\mu\beta(\gamma/k)^2 J_{a,b} - (\gamma/k)^2 K_{a,b}, \qquad (5\text{-}1a)$$

$$\nabla \times H_{a,b} = \gamma^2 \beta \, H_{a,b} - i\omega\varepsilon(\gamma/k)^2 E_{a,b} + i\omega\varepsilon\beta(\gamma/k)^2 K_{a,b} + (\gamma/k)^2 J_{a,b}. \qquad (5\text{-}1b)$$

Using the vector identity $\nabla \cdot [A \times B] = B \cdot [\nabla \times A] - A \cdot [\nabla \times B]$, the twin equations (5-1) can be utilized to obtain

$$\nabla \cdot [E_b \times H_a - E_a \times H_b] =$$
$$(\gamma/k)^2 [i\omega\mu\beta(J_b \cdot H_a - J_a \cdot H_b) + (J_b \cdot E_a - J_a \cdot E_b)$$
$$+ i\omega\varepsilon\beta(K_b \cdot E_a - K_a \cdot E_b) - (K_b \cdot H_a - K_a \cdot H_b)]. \qquad (5\text{-}2)$$

$$i\omega\mu\beta(\gamma/k)^{-2} \nabla \cdot [H_a \times H_b] - i\omega\varepsilon\beta(\gamma/k)^{-2} \nabla \cdot [E_a \times E_b] =$$
$$i\omega\mu\beta(J_a \cdot H_b - J_b \cdot H_a) + i\omega\varepsilon\beta(K_a \cdot E_b - K_b \cdot E_a)$$
$$-k^2\beta^2 (K_a \cdot H_b - K_b \cdot H_a + J_b \cdot E_a - J_a \cdot E_b). \qquad (5\text{-}3)$$

Addition of (5-2) and (5-3) then yields

$$\nabla \cdot [E_b \times H_a - E_a \times H_b + i\omega\mu\beta H_a \times H_b - i\omega\varepsilon\beta E_a \times E_b] =$$
$$(J_b \cdot E_a - K_b \cdot H_a) - (J_a \cdot E_b - K_a \cdot H_b). \qquad (5\text{-}4)$$

Let (5-4) now be integrated over all space. But the volume integral

$$\int_{\text{all space}} dv \ \nabla \cdot [E_b \times H_a - E_a \times H_b + i\omega\mu\beta H_a \times H_b - i\omega\epsilon\beta E_a \times E_b]$$

can be converted into a surface integral over the surface S_∞ bounding all surface by virtue of Gauss' divergence theorem. Using the radiation condition that fields decay as $1/r$ far from their sources, this integral vanishes identically [van Bladel 1985]. Consequently,

$$\int_{\text{all space}} dv \ (J_b \cdot E_a - K_b \cdot H_a)$$
$$= \int_{\text{all space}} dv \ (J_a \cdot E_b - K_a \cdot H_b), \qquad (5\text{-}5)$$

which is akin to the reaction theorem due to Rumsey [1954] for non-chiral media. Therefore, it is clear that *in the application of the reaction theorem an isotropic chiral medium does not differ from an isotropic achiral medium.*

Consideration of (5-2) at a source-free point yields the Lorentz reciprocity theorem [Harrington 1964] for chiral media:

$$\nabla \cdot [E_b \times H_a - E_a \times H_b] = 0. \qquad (5\text{-}6a)$$

Furthermore, by integrating (5-6a) over a sourceless volume V, and the subsequent application of Gauss' divergence theorem, results in

$$\int_S ds \ e_n \cdot [E_b \times H_a - E_a \times H_b] = 0, \qquad (5\text{-}6b)$$

where S is the surface of the source-free volume V, and e_n is the unit outward normal to S. This equation (5-6b) is the Lorentz reciprocity theorem in integral form. Again, *the concept of Lorentz reciprocity is the same for chiral media as for achiral media.* However, yet another Lorentz-type reciprocity equation can be derived from (5-3). The analog of (5-6a) this time is

$$\nabla \cdot [\mu H_a \times H_b - \epsilon E_a \times E_b] = 0, \qquad (5\text{-}7a)$$

which can be restated in integral form as

$$\int_S ds\ e_n \cdot [\mu H_a \times H_b - \varepsilon E_a \times E_b] = 0. \tag{5-7b}$$

Because of the intense use that Babinet's principle [Jackson 1975] finds, it is also of interest to explore the duality relationship of fields radiated by the electric and magnetic sources in that context.

Let $\{E_1, H_1\}$ be the fields produced by electric charge and current densities $\{\rho_e, J\}$, while magnetic charge and current distributions $\{\rho_m, K\}$ independently create the fields $\{E_2, H_2\}$. Maxwell's equations for the fields due to the electric charge and current sources $\{\rho_e, J\}$ read

$$\nabla \times E_1 - i\omega B_1 = 0, \tag{5-8a}$$
$$\nabla \times H_1 + i\omega D_1 = J, \tag{5-8b}$$
$$\nabla \cdot D_1 - \rho_e = 0, \tag{5-8c}$$
$$\nabla \cdot B_1 = 0, \tag{5-8d}$$

while the continuity equation is given by

$$\nabla \cdot J - i\omega \rho_e = 0. \tag{5-8e}$$

Similarly, it can also be shown that

$$\nabla \times H_2 + i\omega D_2 = 0, \tag{5-9a}$$
$$\nabla \times E_2 - i\omega B_2 = -K, \tag{5-9b}$$
$$\nabla \cdot B_2 - \rho_m = 0, \tag{5-9c}$$
$$\nabla \cdot D_2 = 0, \tag{5-9d}$$
$$\nabla \cdot K - i\omega \rho_m = 0. \tag{5-9e}$$

Bearing in mind the constitutive equations (3-5), it is easy to see that the set (5-9) is the dual of the set (5-8), provided the following duality transformations hold:

$$E_1 \leftrightarrow H_2, \qquad\qquad H_1 \leftrightarrow -E_2,$$
$$\mu \leftrightarrow \varepsilon, \qquad\qquad \varepsilon \leftrightarrow \mu, \tag{5-10}$$

$$J \leftrightarrow K, \qquad\qquad \rho_e \leftrightarrow \rho_m,$$
$$\beta \leftrightarrow \beta. \qquad\qquad\qquad\qquad\qquad (5\text{-}10)$$

In other words, *the duality transforms for achiral* [Jackson 1975] *and chiral media turn out to be identical because β remains unchanged during the transformation.*

The reciprocity and the reaction equations derived in this section will be necessary to verify the adequacy and the accuracy of the variously obtained solutions of boundary value and radiation problems. On the other hand, duality relations may come in handy for the simplification of some problems. Although they are commonly held to be fictitious, magnetic sources, K and ρ_m, can be defined so as to convert a difficult problem, -- which involves the electric sources, J and ρ_e, -- into possibly a simpler complementary problem involving magnetic sources. For example, this is commonly done in treatments of scattering by apertures and disks in free space.

6. ENERGY AND MOMENTUM

On an instantaneous basis, the electromagnetic energy conservation law should have the form

$$\nabla \cdot S + \partial W/\partial t + P_{mech} = 0, \tag{6-1}$$

in which the first term, $\nabla \cdot S$, represents the energy flowing out of a given (differential) volume per unit time; $\partial W/\partial t$ is the time rate of change of the stored electromagnetic energy within the same volume; while P_{mech} represents the total work done by the fields on the sources included in that volume. Needless to add, S is the instantaneous Poynting vector. When the medium is non-chiral, i.e., $\beta = 0$, and ε and μ are assumed to be real, then

$$S = E \times H, \tag{6-2a}$$
$$W = (1/2)\,[E \cdot D + H \cdot B] = (1/2)\,[D \cdot D/\varepsilon + B \cdot B/\mu], \tag{6-2b}$$
$$P_{mech} = J \cdot E = J \cdot D/\varepsilon, \tag{6-2c}$$

in which $J = \rho_e\, v$ is an impressed source current density consisting of a charge density ρ_e moving with a velocity v, assuming that there are no magnetic sources.

In order to derive similar expressions for chiral media, one proceeds in the usual fashion [e.g., Jackson 1975] with the latter two Maxwell's equations to obtain

$$\nabla \cdot (E \times H) + H \cdot (\partial B/\partial t) + E \cdot (\partial D/\partial t) + J \cdot E = 0. \tag{6-3}$$

Now, by employing (3-5), assuming that the constitutive parameters are time-invariant, and after some algebraic manipulation [Fedorov 1959a], it can be shown that

$$H \cdot (\partial B/\partial t) \quad = (1/2)\,\{\partial/\partial t\}[H \cdot B] + (1/2)\mu\beta\nabla \cdot [\{\partial H/\partial t\} \times H], \tag{6-4a}$$
$$E \cdot (\partial D/\partial t) \quad = (1/2)\,\{\partial/\partial t\}[E \cdot D] + (1/2)\varepsilon\beta\nabla \cdot [\{\partial E/\partial t\} \times E]. \tag{6-4b}$$

Consequently, (6-3) can be recast as

$$\nabla \cdot \left[\mathbf{E} \times \mathbf{H} - (\beta/2)(\mu\{\partial \mathbf{H}/\partial t\} \times \mathbf{H} + \epsilon\{\partial \mathbf{E}/\partial t\} \times \mathbf{E}) \right]$$
$$+ (\partial/\partial t)\left[(1/2)\{\mathbf{E} \cdot \mathbf{D} + \mathbf{H} \cdot \mathbf{B}\} \right] + \left[\mathbf{J} \cdot \mathbf{E} \right] = 0. \qquad (6\text{-}5a)$$

This equation is identical in form to (6-1), but the interpretation of **S** is different from (6-2a) because of the inclusion of the time-derivatives of **E** and **H**; furthermore, W and P_{mech} now have the same definitions as the first set of definitions in (6-2b) and (6-2c), respectively. Further manipulation of (6-5a), however, leads to the elimination of the time-derivatives on which the divergence also operates, and yields [Fedorov 1959b]

$$\nabla \cdot \left[\mathbf{E} \times \mathbf{H} \right] + (\partial/\partial t)\left[(1/2)\{\mathbf{D} \cdot \mathbf{D}/\epsilon + \mathbf{B} \cdot \mathbf{B}/\mu\} \right] + \left[\mathbf{J} \cdot \mathbf{D}/\epsilon \right] = 0. \qquad (6\text{-}5b)$$

In this second expression, it should be noted that the form of **S** has been restored to that in (6-2a); however, W and P_{mech} now have definitions the same as the second set of definitions in (6-2b) and (6-2c), respectively. Whereas the two definitions of S, W and P_{mech} from (6-5a) and (6-5b) are identical if $\beta = 0$, they are not obviously so for isotropic chiral media. Interestingly enough, quantities of the form $\{\partial \mathbf{E}/\partial t\} \times \mathbf{E}$ are components of what has been called the zilch tensor, obtained by considering some non-standard conservation laws for the electromagnetic field in vacuum [Lipkin 1964; Kibble 1965; Ragusa 1988].

For time-harmonic fields, the conservation of energy principle can be similarly obtained. Corresponding to (6-5a), this principle can be stated as

$$0 = \nabla \cdot \mathbf{P} + (1/2)\, Re\{i\omega(\mathbf{E} \cdot \mathbf{D}^* - \mathbf{H} \cdot \mathbf{B}^*)\} + (1/2)\, Re\{\mathbf{E} \cdot \mathbf{J}^*\}, \qquad (6\text{-}6)$$

while, corresponding to (6-5b), it is given as

$$0 = \nabla \cdot \mathbf{P} + (1/2)\, Re\{i\omega(\mathbf{D} \cdot \mathbf{D}^*/\epsilon - \mathbf{B} \cdot \mathbf{B}^*/\mu)\} + (1/2)\, Re\{\mathbf{D} \cdot \mathbf{J}^*/\epsilon\}, \qquad (6\text{-}7)$$

in which

$$\mathbf{P} = (1/2)Re\{\mathbf{E} \times \mathbf{H}^*\} \qquad (6\text{-}8)$$

is the usual time-averaged time-harmonic Poynting vector, with the asterisk denoting the complex conjugate and *Re* meaning the 'real part of.'

Finally in this section, the conservation of electromagnetic momentum [Jackson 1975] is considered for the Drude-Born-Fedorov media. Utilizing Maxwell's equations and the constitutive relations (3-5) it is easy to show that

$$\mathbf{J}\times\mathbf{B} + (\nabla\cdot\mathbf{D})\mathbf{E} + (\partial/\partial t)[\mathbf{D}\times\mathbf{B}]$$
$$= \varepsilon[\mathbf{E}(\nabla\cdot\mathbf{E}) - \mathbf{E}\times(\nabla\times\mathbf{E})] + \mu[\mathbf{H}(\nabla\cdot\mathbf{H}) - \mathbf{H}\times(\nabla\times\mathbf{H})]. \quad (6\text{-}9)$$

But $\nabla\cdot\mathbf{D} = \rho_e$, the charge density, while $\mathbf{J} = \rho_e \mathbf{v}$. Hence,

$$\mathbf{J}\times\mathbf{B} + (\nabla\cdot\mathbf{D})\mathbf{E} = \rho_e[\mathbf{E} + \mathbf{v}\times\mathbf{B}], \quad (6\text{-}10)$$

which is an expression of the Lorentz force. Consequently, any volume integral of the left side of (6-10) must be interpreted as the time-derivative of the sum of the *mechanical* momenta of all charges included in that volume, i.e.,

$$(\partial/\partial t)\, \mathbf{G}_{mech} = \int_V dv\, \rho_e[\mathbf{E} + \mathbf{v}\times\mathbf{B}]. \quad (6\text{-}11)$$

This immediately suggests that the volume integral of $(\partial/\partial t)[\mathbf{D}\times\mathbf{B}]$ in (6-9) must be the *electromagnetic* momentum $\mathbf{G}_{em}$ as per

$$\mathbf{G}_{em} = \int_V dv\, [\mathbf{D}\times\mathbf{B}]. \quad (6\text{-}12)$$

Finally, the right side of (6-9) can be used to define the Maxwell stress tensor $\mathfrak{T}$ as applicable to chiral media. The cartesian components of $\mathfrak{T}$ are defined as

$$(\mathfrak{T})_{nm} = \varepsilon[E_n E_m - (1/2)\, E_n E_n\, \delta_{nm}]$$
$$+ \mu[H_n H_m - (1/2)\, H_n H_n\, \delta_{nm}]; \quad n,m = 1,2,3, \quad (6\text{-}13)$$

with δ_{nm} being the Kronecker delta. In terms of the definitions (6-11) - (6-13), the principle of conservation of momentum can then be compactly set down as

28

$$(\partial/\partial t) \, [\mathbf{G}_{mech} + \mathbf{G}_{em}]_n$$
$$= \sum_{m=1,2,3} \int_V dv \ (\partial/\partial x_m) \, (\mathfrak{T})_{nm}; \ n = 1,2,3. \qquad (6\text{-}14)$$

7. BOHREN'S DECOMPOSITION

Generally speaking, only finite volumes may be endowed with chirality; and until the development of composite media, there was no specific need to develop a full-fledged electromagnetic theory pertinent to chiral media. Furthermore, native chirality is exhibited by organic substances only at optical frequencies, where ray-tracing techniques suffice for the interpretation of experimentally obtained data. Thus, it was only in the early 1970's that the need arose for more general treatments. In this context, Bohren [1974,1975] devised a transformation which permits the solution of such problems as scattering by chiral spheres.

Some background on the polarization state of a plane wave in an achiral, isotropic, homogeneous medium may be necessary to appreciate Bohren's transformation. The instantaneous electric field is real-valued, and can be calculated as $Re\{E \exp[-i\omega t]\}$. The complex vector E of any homogeneous plane wave is of the form $A \exp[i\sigma e_\eta \cdot r]$, in which A is a complex amplitude, e_η is a unit vector in the direction of propagation, and σ is the associated (real) wavenumber. A general homogeneous plane wave is *elliptically* polarized since the curve traced out by the tip of $Re\{A \exp[i(\sigma e_\eta \cdot r - \omega t)]\}$, at a fixed position r and as a function of time t, is an ellipse. If $ie_\eta \cdot (A \times A^*) = 0$, then the ellipse degenerates into a straight line, and the plane wave is said to be *linearly* polarized. A right-handed elliptically polarized plane wave satisfies the condition $ie_\eta \cdot (A \times A^*) > 0$, while a left-handed one obeys the inequality $ie_\eta \cdot (A \times A^*) < 0$. More pertinently in the present context, a right-circularly polarized (RCP) plane wave must conform to the prescription $A = ie_\eta \times A^*$, whereas a left-circularly polarized (LCP) plane wave has $A = -ie_\eta \times A^*$.

The essence of Bohren's transformation is the prescription of a LCP field Q_1 and a RCP field Q_2 in a chiral medium, defined as per the following relations:

$$E = Q_1 + a_R \ Q_2, \tag{7-1a}$$
$$H = a_L \ Q_1 + Q_2, \tag{7-1b}$$

with

$$a_R = -i \ \sqrt{(\mu/\epsilon)} = -1/a_L. \tag{7-2}$$

In these equations, a_R carries the unit of an impedance and a_L, that of an admittance. In an unbounded chiral medium LCP and RCP fields can exist independently. From Bohren's transformation, it follows that $\mathbf{E} = i\sqrt{(\mu/\varepsilon)}\mathbf{H}$ for a LCP field, and $\mathbf{E} = -i\sqrt{(\mu/\varepsilon)}\mathbf{H}$ for a RCP field; consequently, the ratio $\sqrt{(\mu/\varepsilon)}$ can be considered as the intrinsic wave impedance of a chiral medium, just as in the case of achiral media.

Substitution of (7-1) into source-free Maxwell's equations leads to the wave equations

$$\nabla^2\,\mathbf{Q}_1 + \gamma_1{}^2\,\mathbf{Q}_1 = 0, \qquad\qquad \nabla^2\,\mathbf{Q}_2 + \gamma_2{}^2\,\mathbf{Q}_2 = 0, \qquad\qquad (7\text{-}3)$$

with the respective wavenumbers

$$\gamma_1 = k/(1\text{-}k\beta), \qquad\qquad\qquad \gamma_2 = k/(1\text{+}k\beta). \qquad\qquad (7\text{-}4)$$

The polarization characteristics of the $\mathbf{Q}$ fields are reaffirmed via the circulation equations [Chen 1983]

$$\nabla\times\mathbf{Q}_1 = \gamma_1\,\mathbf{Q}_1, \qquad\qquad\qquad \nabla\times\mathbf{Q}_2 = -\,\gamma_2\,\mathbf{Q}_2; \qquad\qquad (7\text{-}5)$$

in addition, both the $\mathbf{Q}$ fields are divergenceless.

Bohren's decomposition shows that all fields in a sourceless chiral volume are either LCP or RCP; it is left for later sections to show that radiating sources can emit only circularly polarized fields in isotropic chiral media. The admissible planewave solutions of (4-7) are, respectively, left-circularly polarized (LCP) and right-circularly polarized (RCP), and are given as

$$\mathbf{Q}_1 = (\mathbf{e}_{\parallel} + i\mathbf{e}_{\perp})\,\exp[i\gamma_1\,\mathbf{e}_p{\cdot}\mathbf{r}], \qquad\qquad\qquad (7\text{-}6)$$
$$\mathbf{Q}_2 = (\mathbf{e}_{\parallel} - i\mathbf{e}_{\perp})\,\exp[i\gamma_2\,\mathbf{e}_p{\cdot}\mathbf{r}\,], \qquad\qquad\qquad (7\text{-}7)$$

in which $\mathbf{e}_p$ is the direction of propagation; $\mathbf{e}_p$, $\mathbf{e}_{\parallel}$ and $\mathbf{e}_{\perp}$ are mutually orthogonal unit vectors forming a cartesian co-ordinate system.

If β is assumed to have a positive real part (right-handed medium), then $\mathbf{Q}_1$ propagates with a slower phase velocity; or else, $\mathbf{Q}_2$ is the slower of the two waves. In addition, in unbounded, isotropic chiral media these planewaves are transverse electromagnetic (TEM). It must, however, be emphasized here that combining Q_1 and Q_2 in any fashion cannot lead to a linearly polarized planewave, unless $\beta = 0$, because the LCP and the RCP plane waves travel with different phase velocities.

In an unbounded chiral medium, the LCP and the RCP plane waves can propagate without interfering with each other. But when a wave of either polarization encounters a boundary, mode conversion takes place; the scattered field then consists, in general, of waves of both (circular) polarizations. Unattenuated propagation of both the LCP and the RCP waves occurs provided both k and β are real. If $Im(k) + |k|^2 \, Im(\beta) = 0$, γ_1 is real and Q_1 traverses the chiral medium without losing any energy, with Im meaning the 'imaginary part of.' On the other hand, if $Im(k) - |k|^2 \, Im(\beta) = 0$, γ_2 is real and Q_2 propagates through the medium without suffering any attenuation.

The time-averaged Poynting vectors for the fields Q_1 and Q_2 can be easily derived using the decompositions (7-1) in the definition (6-8). Consequently,

$$P_1 = (1/2) \, Re \, \{Q_1 \times a_L{}^* Q_1{}^*\}, \qquad\qquad (7\text{-}8a)$$
$$P_2 = (1/2) \, Re \, \{a_R \, Q_2 \times Q_2{}^*\}, \qquad\qquad (7\text{-}8b)$$

while the cross-coupling of Q_1 and Q_2 cannot give rise to any energy since $\gamma_1 \neq \gamma_2$. Likewise, the Rumsey reaction theorem (5-5) for these circularly-polarized fields can be obtained as

$$\int_{\text{all space}} dv \, \{Q_{1a} \cdot [J_b + (i\omega\varepsilon/k)K_b] - Q_{2a} \cdot [(i\omega\mu/k)J_b + K_b]\} =$$
$$\int_{\text{all space}} dv \, \{Q_{1b} \cdot [J_a + (i\omega\varepsilon/k)K_a] - Q_{2b} \cdot [(i\omega\mu/k)J_a + K_a]\}. \qquad (7\text{-}9)$$

In the remainder of this work, primary emphasis will be on the E and the H fields, but Q_1 and Q_2 will be used whenever it is advantageous to do so. In general, the use of E and H is warranted in the solution of boundary value problems because the boundary conditions are given in terms of these fields. On the other hand, radiation as well as propagation problems are considerably simplified by the use of Q_1 and Q_2. It is amusing to reflect upon the fact that while E and H are incapable of independent propagation but decouple at the boundaries, Q_1 and Q_2 propagate independently but coupled at boundaries.

To be noted are the relations

$$\beta = [\gamma_2^{-1} - \gamma_1^{-1}]/2, \qquad\qquad k^{-1} = [\gamma_2^{-1} + \gamma_1^{-1}]/2, \qquad\qquad (7\text{-}10)$$

which, together with the immittances (7-2), express the constitutive parameters in terms of the wavenumbers γ_1 and γ_2. Let a linearly polarized plane wave be normally incident on a planar vacuum/chiral interface. If the (lossless) chiral medium is a slab of thickness d, the plane wave transmitted into free space on the other side has its plane of polarization rotated by an angle

$$\delta = \pi \, d \, (\gamma_1 - \gamma_2). \qquad\qquad (7\text{-}11)$$

Measurements of this rotation, called optical rotatory dispersion (ORD) abound, -- as also are measurements, called circular dichroism (CD) of the differential absorption of LCP and RCP plane waves; -- see, e.g., Charney [1979]. But measurements of γ_1 and γ_2 individually are rarely made, the only example known to the authors being that for poly-L-glutamic acid in the $1.25\text{-}1.58\times10^{15}$ Hz frequency range [Urry & Krivacic 1970]; for this material, $k\beta$ is of the order 10^{-4} and smaller in magnitude, while $|\beta|$ is of the order 10^{-12} and smaller. Thus, experimental confirmation of the theoretical results of the following sections must await the measurment of ε, μ and β of chiral media. There is evidence that, with interest in constructing artificial chiral composites, active at millimeter-wave frequencies, these properties will become available in the near future [Guire et al 1988].

8. REFLECTION AND TRANSMISSION OF PLANE WAVES

Layered structures are often the easiest to analyze, and for that purpose the reflection and transmission characteristics of a planar achiral/chiral interface serve as important building blocks. Let the region $z \geq 0$ be occupied by a Drude-Born-Fedorov chiral medium, while the region $z \leq 0$ is assumed to be free space, as shown in Fig. 8.1. A homogeneous plane wave of arbitrary polarization is incident upon the bimaterial interface $z = 0$ from the free space side, and is given by

$$\mathbf{E}_{inc} = [A_E \, \mathbf{e}_y - A_H \, (\alpha_o/k_o) \, \mathbf{e}_x + A_H \, (\kappa/k_o)\mathbf{e}_z] \, \exp[i(\kappa x + \alpha_o z)], \qquad (8\text{-}1a)$$

$$\mathbf{H}_{inc} = (k_o/\omega\mu_o) \, [-A_H \, \mathbf{e}_y - A_E \, (\alpha_o/k_o) \, \mathbf{e}_x + A_E \, (\kappa/k_o)\mathbf{e}_z]$$
$$\exp[i(\kappa x + \alpha_o z)]. \qquad (8\text{-}1b)$$

In these expressions, $k_o = \omega\sqrt{(\mu_o\varepsilon_o)}$ is the free space wavenumber; $\alpha_o = +\sqrt{\{k_o^2 - \kappa^2\}}$; while, if θ_o is the (real) angle made by the propagation vector of the incident plane wave with the $+z$ axis, then $\kappa/k_o = \sin\theta_o$. Whereas the coefficients $A_E \neq 0$, $A_H = 0$ refer to a TE-polarized incident planewave, $A_E = 0$, $A_H \neq 0$ denote an incident TM-polarized field. Reflection and transmission characteristics for circularly polarized plane waves can also be studied: $A_H = i \, A_E$ for LCP and $A_H = -i \, A_E$ for RCP plane waves.

In the chiral region, the existing field has to be expressed in terms of the LCP and the RCP plane waves. Consistent with Snell's laws, the appropriate representation of the transmitted field is given by

$$Q_1 = A_1 \, [\mathbf{e}_y + i(-\mathbf{e}_x \, \alpha_1 + \mathbf{e}_z \, \kappa)/\gamma_1] \, \exp[i(\kappa x + \alpha_1 z)]; \quad z \geq 0, \qquad (8\text{-}2a)$$

$$Q_2 = A_2 \, [\mathbf{e}_y + i(\mathbf{e}_x \, \alpha_2 - \mathbf{e}_z \, \kappa)/\gamma_2] \, \exp[i(\kappa x + \alpha_2 z)]; \quad z \geq 0. \qquad (8\text{-}2b)$$

In this representation, the coefficients A_1 and A_2 are unknowns to be determined from the solution of the boundary value problem; and the transmitted field in the chiral half space is given by

$$\mathbf{E}_{trans} = Q_1 + a_R \, Q_2, \qquad \mathbf{H}_{trans} = a_L \, Q_1 + Q_2, \qquad (8\text{-}3)$$

as per Bohren's decomposition. The parameters $\alpha_1 = +\sqrt{\{\gamma_1^2 - \kappa^2\}}$ and $\alpha_2 = +\sqrt{\{\gamma_2^2 - \kappa^2\}}$, while a_R and a_L have been defined by (7-2).

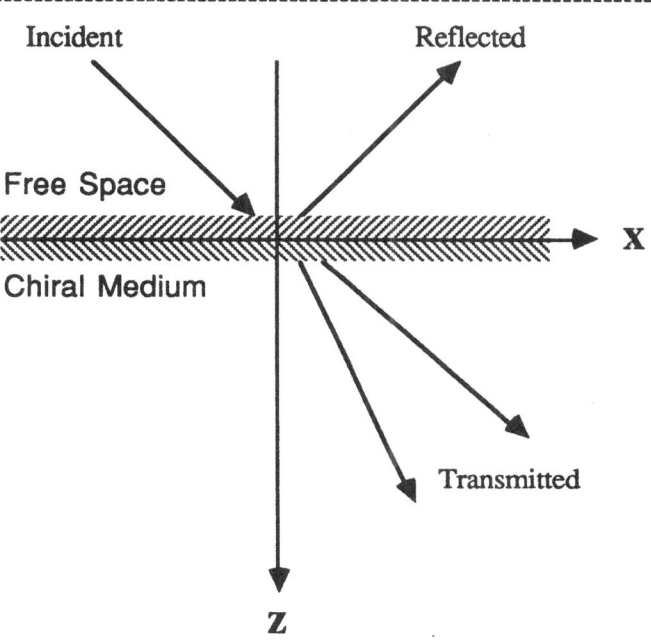

Fig. 8-1. Planewave reflection and transmission characteristics of an achiral/chiral interface.

Because of the circular birefringence in the chiral half space, the reflected field will have both TE- and TM- polarized plane wave components. It will be best, therefore, if the reflected field in the free space region ($z \leq 0$) is also expressed in terms of circularly polarized plane waves:

$$\mathbf{E}_{ref} = \{ C_1 [e_y + i(e_x\alpha_o + e_z\kappa)/k_o]$$
$$+ C_2[e_y - i(e_x\alpha_o + e_z\kappa)/k_o] \} \exp[i(\kappa x - \alpha_o z)]; \; z \leq 0, \quad (8\text{-}4a)$$
$$\mathbf{H}_{ref} = (k_o/i\omega\mu_o) \{ C_1 [e_y + i(e_x\alpha_o + e_z\kappa)/k_o]$$
$$- C_2[e_y - i(e_x\alpha_o + e_z\kappa)/k_o] \} \exp[i(\kappa x - \alpha_o z)]; \; z \leq 0, \quad (8\text{-}4b)$$

in which the coefficients C_1 and C_2 have to be determined as well.

The solution to the problem is straightforward: the satisfaction of Snell's laws has already been built into the field representations via κ, while the usual boundary conditions prescribed on the tangential components of $\mathbf{E}$ and $\mathbf{H}$ fields,

$$e_x \cdot [E_{inc} + E_{ref}] = e_x \cdot E_{trans}; \quad z = 0, \tag{8-5a}$$

$$e_y \cdot [E_{inc} + E_{ref}] = e_y \cdot E_{trans}; \quad z = 0, \tag{8-5b}$$

$$e_x \cdot [H_{inc} + H_{ref}] = e_x \cdot H_{trans}; \quad z = 0, \tag{8-5c}$$

$$e_y \cdot [H_{inc} + H_{ref}] = e_y \cdot H_{trans}; \quad z = 0, \tag{8-5d}$$

lead to the solution of a 4×4 matrix equation from which the unknown coefficients of the reflected and the transmitted plane waves can be determined. In the special case of normal incidence (i.e., $\kappa = 0$), the reflection and the transmission coefficients can be easily obtained as

$$C_1 = (1/2)(A_E + iA_H) \, [\sqrt{(\mu/\mu_o)} - \sqrt{(\varepsilon/\varepsilon_o)}][\sqrt{(\mu/\mu_o)} + \sqrt{(\varepsilon/\varepsilon_o)}]^{-1}, \tag{8-6a}$$

$$C_2 = (1/2)(A_E - iA_H) \, [\sqrt{(\mu/\mu_o)} - \sqrt{(\varepsilon/\varepsilon_o)}][\sqrt{(\mu/\mu_o)} + \sqrt{(\varepsilon/\varepsilon_o)}]^{-1}, \tag{8-6b}$$

$$A_1 = (1/2)(A_E - iA_H) \, [\sqrt{(\mu/\mu_o)}][\sqrt{(\mu/\mu_o)} + \sqrt{(\varepsilon/\varepsilon_o)}]^{-1}, \tag{8-6c}$$

$$A_2 = (1/2)(A_E + iA_H) \, [\sqrt{(\varepsilon/\mu_o)}][\sqrt{(\mu/\mu_o)} + \sqrt{(\varepsilon/\varepsilon_o)}]^{-1}. \tag{8-6d}$$

These results confirm that a normally incident LCP (resp. RCP) plane wave is reflected as a RCP (resp. LCP) plane wave, while the transmitted plane wave is LCP (resp. RCP).

In modern times this problem has been systematically studied by Ramachandran and Ramaseshan [1961], as well as by Bokut' and Fedorov [1960], and more recently by Lakhtakia et al [1986] and Bassiri et al [1988]. Extension to the absorption characteristics of a metal-backed chiral slab has also been made by Varadan et al [1987a], and the response of a chiral slab suspended in free space has been studied by Bokut' and Sotskii [1963]. Irradiation of a planar achiral/chiral interface by the near field of a line source has also been examined [Lakhtakia et al 1988d].

9. THE IMAGING CONCEPT

With the developments relating to the response of layered planar geometries, it is time to investigate the concept of imaging which is so central to diverse problems of practical interest relating to achiral media. For this purpose, let two problems be considered.

In *Problem 1*, let the space $z \leq 0$ be occupied by the chiral medium $\{\varepsilon,\mu,\beta\}$, while the half-space $z \geq 0$ is occupied by the mirror-conjugate medium $\{\varepsilon,\mu,-\beta\}$. Either a LCP or a RCP plane wave is incident on the interface $z = 0$ from the zone $z \leq 0$. It will be appropriate to express the fields in the zone $z \leq 0$ by the fields

$$Q_1 = A_1 \, [e_y + i(-e_x \, \alpha_1 + e_z \, \kappa)/\gamma_1] \, \exp[i(\kappa x + \alpha_1 z)] \, +$$
$$+ B_1 \, [e_y + i(e_x \, \alpha_1 + e_z \, \kappa)/\gamma_1] \, \exp[i(\kappa x - \alpha_1 z)]; \quad z \leq 0, \qquad (9\text{-}1a)$$
$$Q_2 = A_2 \, [e_y + i(e_x \, \alpha_2 - e_z \, \kappa)/\gamma_2] \, \exp[i(\kappa x + \alpha_2 z)] \, +$$
$$+ B_2 \, [e_y - i(e_x \, \alpha_2 + e_z \, \kappa)/\gamma_2] \, \exp[i(\kappa x - \alpha_2 z)]; \quad z \leq 0. \qquad (9\text{-}1b)$$

In these equations, A_1 and A_2 represent the incident plane waves, while B_1 and B_2 are the amplitudes of the reflected ones; κ is the horizontal wavenumber required by Snell's laws; and as before, $\alpha_1 = +\sqrt{\{\gamma_1{}^2 - \kappa^2\}}$ and $\alpha_2 = +\sqrt{\{\gamma_2{}^2 - \kappa^2\}}$; while e_x, etc., are the unit cartesian vectors.

The half-space $z \geq 0$ is occupied by the mirror-conjugate medium; this means that phase velocities of the LCP and the RCP plane waves here are, respectively, those of the RCP and the LCP plane waves in the medium of incidence. Consequently, an acceptable representation of the fields in the medium of transmission is given by

$$Q_1 = C_1 \, [e_y + i(-e_x \, \alpha_2 + e_z \, \kappa)/\gamma_2] \, \exp[i(\kappa x + \alpha_2 z)]; \quad z \geq 0, \qquad (9\text{-}2a)$$
$$Q_2 = C_2 \, [e_y + i(e_x \, \alpha_1 - e_z \, \kappa)/\gamma_1] \, \exp[i(\kappa x + \alpha_1 z)]; \quad z \geq 0, \qquad (9\text{-}2b)$$

with C_1 and C_2 being the transmission coefficients.

The solution of the boundary value problem is sought in the convenient matrix forms

$$\begin{pmatrix} B_1 \\ B_2 \end{pmatrix} = \begin{pmatrix} R_{11} & R_{12} \\ R_{21} & R_{22} \end{pmatrix} \begin{pmatrix} A_1 \\ A_2 \end{pmatrix} \quad , \tag{9-3a}$$

$$\begin{pmatrix} C_1 \\ C_2 \end{pmatrix} = \begin{pmatrix} T_{11} & T_{12} \\ T_{21} & T_{22} \end{pmatrix} \begin{pmatrix} A_1 \\ A_2 \end{pmatrix} \quad , \tag{9-3b}$$

with the R's constituting the reflection matrix; and the T's, the transmission matrix. The utilization of (9-1) and (9-2) in ensuring that there are no discontinuities in the tangential E- and the tangential H- fields across the interface $z = 0$ yields:

$$\begin{aligned} R_{11} &= - R_{22} = (\alpha_1 \gamma_2 - \alpha_2 \gamma_1)/ (\alpha_1 \gamma_2 + \alpha_2 \gamma_1), \\ T_{11} &= 2 \alpha_1 \gamma_2 / (\alpha_1 \gamma_2 + \alpha_2 \gamma_1), \\ T_{22} &= 2 \alpha_2 \gamma_1/ (\alpha_1 \gamma_2 + \alpha_2 \gamma_1), \\ R_{12} &= R_{21} = T_{12} = T_{21} = 0, \end{aligned} \tag{9-4}$$

where a_R and a_L have been defined by (7-2).

This curious result should be noted : if the incident plane wave is LCP (resp. RCP), then the reflected as well as the transmitted waves are also LCP (resp. RCP). There are no waves of the opposite handedness generated at the planar boundaries between mirror-conjugated chiral media. Thus, the arrangement of Problem 1 acts somewhat like a beam-splitter; an incident LCP (resp. RCP) plane wave is broken up into two LCP (resp. RCP) plane waves, each of which leaves the interface in opposite z-directions with different amplitudes and phase velocities. Parenthetically, it is mentioned here that the foregoing result is not a consequence of the two media being mirror-conjugated; rather, as is proved by the analysis leading to (17-17) and (17-18), it is due to the fact that the mirror-conjugated media are impedance-matched.

In *Problem 2*, let the medium in the zone $z \geq 0$ be perfectly conducting. Then, a solution of the form

38

Fig. 9-1: Illustrating the correspondence between Problems 1 and 2 as per the relations (9-7).

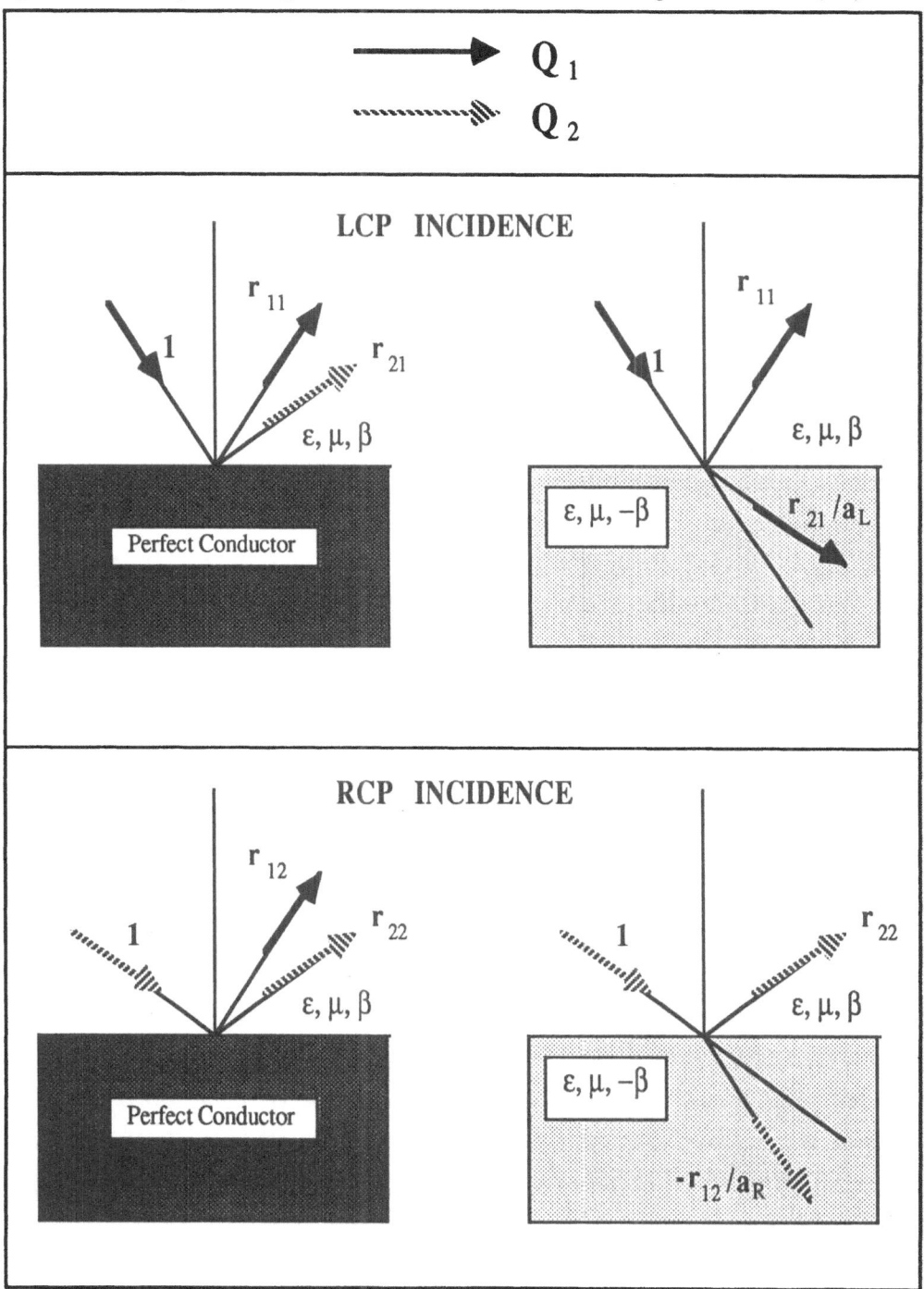

$$\begin{pmatrix} B_1 \\ B_2 \end{pmatrix} = \begin{pmatrix} r_{11} & r_{12} \\ r_{21} & r_{22} \end{pmatrix} \begin{pmatrix} A_1 \\ A_2 \end{pmatrix} \qquad (9\text{-}5)$$

is to be sought. For this purpose, (9-1) are utilized to ensure the nulling of the tangential E-field at the impenetrable surface $z = 0$, and the result can be obtained as

$$r_{11} = -r_{22} = (\alpha_1 \gamma_2 - \alpha_2 \gamma_1)/(\alpha_1 \gamma_2 + \alpha_2 \gamma_1)$$
$$r_{21} = 2 a_L \alpha_1 \gamma_2 /(\alpha_1 \gamma_2 + \alpha_2 \gamma_1)$$
$$r_{12} = -2 a_R \alpha_2 \gamma_1 /(\alpha_1 \gamma_2 + \alpha_2 \gamma_1). \qquad (9\text{-}6)$$

It is to be observed, therefore, that when a LCP or a RCP plane wave hits a perfectly conducting surface, then the reflected field consists of components of both circular polarization states. However, when $\kappa = \gamma_1$, then $r_{21} = 0$, $r_{11} = -1$, $r_{12} = -2a_R$ and $r_{22} = 1$ giving rise to the trivial case of no reflection for the grazing LCP incidence. Likewise, when $\kappa = \gamma_2$, then $r_{12} = 0$, $r_{22} = -1$, $r_{21} = 2a_L$ and $r_{11} = 1$.

There is a correspondence between the solutions of Problems 1 and 2, respectively; that is

$$R_{11} = r_{11}, \qquad\qquad R_{22} = r_{22}$$
$$T_{11} = r_{21}/a_L \qquad\qquad T_{22} = -r_{12}/a_R . \qquad (9\text{-}7)$$

For the sake of illustration, let us consider the case of LCP incidence. In both Problems 1 and 2, r_{11} is the amplitude of the reflected LCP wave which travels with a phase velocity ω/γ_1. In Problem 2, r_{21} is the amplitude of the *reflected RCP* wave with a phase velocity ω/γ_2; but in Problem 1, r_{21}/a_L is the amplitude of the *transmitted LCP* wave which also travels with a phase velocity ω/γ_2 because the medium of transmission is the mirror-conjugate of the medium of incidence and reflection. Analogous comments also apply to the case of RCP plane wave incidence, and both cases have been schematically illustrated in Fig. 9-1. It should be noted that Fermat's principle is equally well satisfied in both Problems 1 and 2, and in identical fashion.

Fig. 9-2 : To illustrate the imaging concept for Problem 2 by using two specializations of Problem 1.

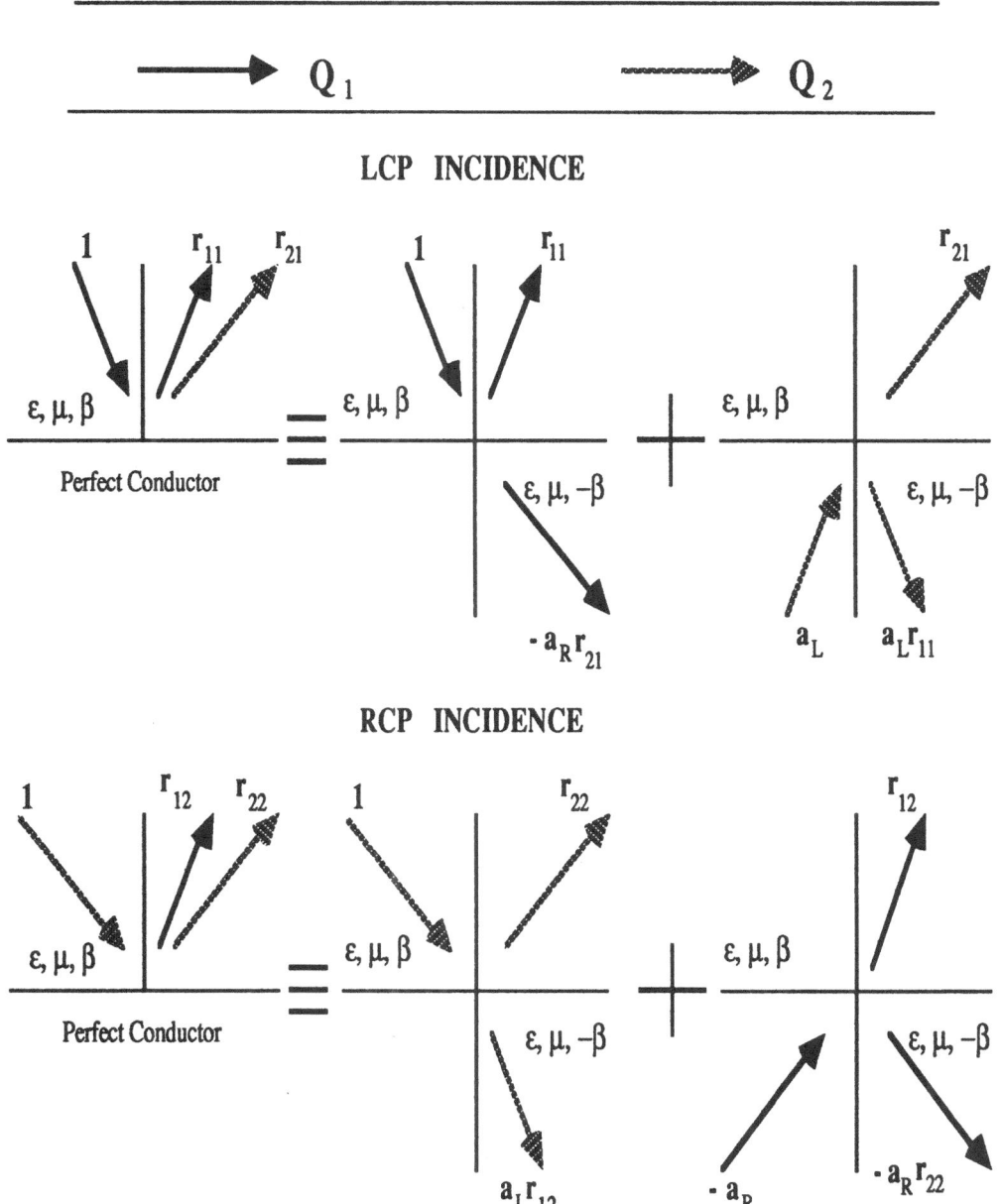

The correspondences given in (9-7), coupled with the complementary relations

$$r_{LL}(-\beta) = r_{RR}(\beta), \qquad\qquad r_{RR}(-\beta) = r_{LL}(\beta),$$
$$r_{LR}(-\beta) = a_R{}^2 r_{RL}(\beta), \qquad\qquad r_{RL}(-\beta) = a_L{}^2 r_{LR}(\beta), \qquad\qquad (9\text{-}8)$$

point out to the imaging concept. Consider Problem 2 again with an incident LCP plane wave of unit amplitude. Relevant to the zone $z \leq 0$, this problem is equivalent to the superposition of two problems, each of which is a specialization of Problem 1. These two problems are: (i) a Problem 1 in which a LCP plane wave is incident on the interface from the zone $z \leq 0$ with unit amplitude, and (ii) a Problem 1 in which a RCP plane wave is incident on the interface from the zone $z \geq 0$ with an amplitude equal to a_L. The case of an incident RCP plane wave in Problem 2 can also be handled thus, and both cases are schematically illustrated in Fig. 9-2. But, as is clear from Fig. 9-2, the use of an imaging theory for chiral media is very complicated for scattering problems in general: not only do the sources get imaged, the medium also does.

10. SCATTERING BY A CIRCULAR CHIRAL CYLINDER

A problem of considerable interest concerns the scattering response of chiral cylinders which are infinitely long and have a right circular cross-section. Bohren [1978] has examined the scattering of plane waves by chiral cylinders, and his treatment is recapitulated in brief here. As is customary in two-dimensional problems, two mutually exclusive situations will be analyzed: the TE-case in which the incident electric field is polarized parallel to the cylindrical axis (i.e., the z axis); and the TM- case in which the incident magnetic field is polarized parallel to the z axis. The two situations are illustrated in Fig. 10-1. The cylinder is considered to be embedded in free space, and $k_0 = \omega\sqrt{(\mu_0\varepsilon_0})$ is the free space wavenumber.

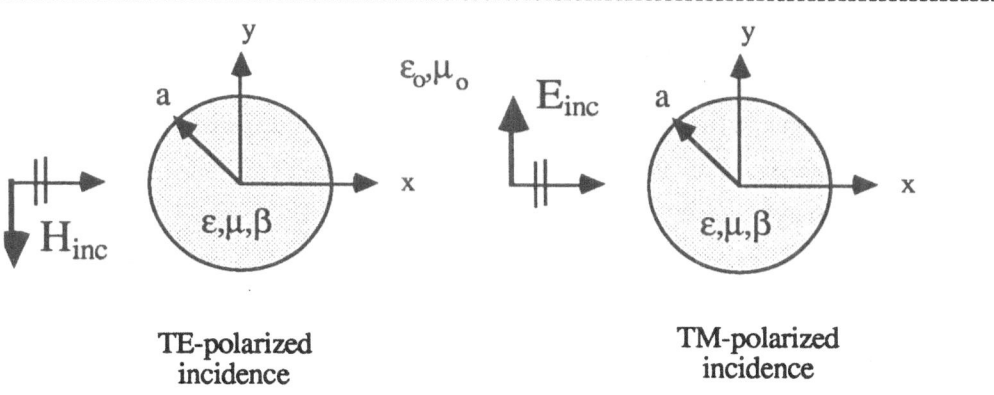

TE-polarized incidence

TM-polarized incidence

Fig. 10-1. Pertinent to the scattering of TE- and TM- polarized plane waves by a chiral cylinder.

Let the transverse electric (TE) case be considered first, i.e., the incident electric field is given by $E_{inc} = e_z\exp[ik_0x]$. Then, the incident electromagnetic field can be expanded in terms of vector functions

$$E_{inc} = \sum_{n \in [-\infty,\infty]} (i^n/k_0) \, N_n^{(1)}(k_0r) = e_z\exp[ik_0x], \qquad (10\text{-}1a)$$
$$H_{inc} = (k_0/i\omega\mu_0) \sum_{n \in [-\infty,\infty]} (i^n/k_0) \, M_n^{(1)}(k_0r), \qquad (10\text{-}1b)$$

in which

$$\mathbf{M}_n^{(1)}(\sigma\mathbf{r}) = \nabla\times[\mathbf{e}_z \exp[in\varphi]J_n(\sigma r)], \tag{10-2a}$$

$$\mathbf{N}_n^{(1)}(\sigma\mathbf{r}) = (1/\sigma)\nabla\times\nabla\times[\mathbf{e}_z \exp[in\varphi]J_n(\sigma r)], \tag{10-2b}$$

with $J_n(\sigma r)$ being the cylindrical Bessel function of order n, $\mathbf{r} = x\mathbf{e}_x+y\mathbf{e}_y$ and $\varphi = \arctan(y/x)$. The scattered field is similarly expanded:

$$\mathbf{E}_{sc} = -\sum_{n\in[-\infty,\infty]} (i^n/k_o)\{ic_n\mathbf{M}_n^{(3)}(k_o\mathbf{r}) + b_n\mathbf{N}_n^{(3)}(k_o\mathbf{r})\}; \quad r \geq a, \tag{10-3a}$$

$$\mathbf{H}_{sc} = (k_o/i\omega\mu_o)\sum_{n\in[-\infty,\infty]} (i^n/k_o)\{ic_n\mathbf{N}_n^{(3)}(k_o\mathbf{r}) + b_n\mathbf{M}_n^{(3)}(k_o\mathbf{r})\}, \tag{10-3b}$$

$$\mathbf{M}_n^{(3)}(\sigma\mathbf{r}) = \nabla\times[\mathbf{e}_z \exp[in\varphi]H_n^{(1)}(\sigma r)], \tag{10-4a}$$

$$\mathbf{N}_n^{(3)}(\sigma\mathbf{r}) = (1/\sigma)\nabla\times\nabla\times[\mathbf{e}_z \exp[in\varphi]H_n^{(1)}(\sigma r)], \tag{10-4b}$$

with $H_n^{(1)}(\sigma r)$ being the cylindrical Hankel function of the first kind and order n; b_n and c_n are the unknown coefficients of expansion. The field expansions (10-3) are valid outside the boundary $r = a$ of the cylinder, and also conform to the radiation conditions. Finally, *vide* Bohren's decomposition, the field inside the chiral cylinder is given by

$$\mathbf{Q}_1 = \sum_{n\in[-\infty,\infty]} (i^n/k_o)g_n\{\mathbf{M}_n^{(1)}(\gamma_1\mathbf{r}) + \mathbf{N}_n^{(1)}(\gamma_1\mathbf{r})\}; \quad r \leq a \tag{10-5a}$$

$$\mathbf{Q}_2 = \sum_{n\in[-\infty,\infty]} (i^n/k_o)f_n\{\mathbf{M}_n^{(1)}(\gamma_2\mathbf{r}) - \mathbf{N}_n^{(1)}(\gamma_2\mathbf{r})\}; \quad r \leq a \tag{10-5b}$$

The solution requires the use of the boundary conditions that there be no discontinuities in the tangential $\mathbf{E}$ and $\mathbf{H}$ fields at $r = a$; consequently,

$$b_n = b_{-n} = [W_{n2}B_{n1}+W_{n1}B_{n2}]/[W_{n2}V_{n1}+W_{n1}V_{n2}], \tag{10-6a}$$

$$c_n = c_{-n} = i[V_{n2}B_{n1}-V_{n1}B_{n2}]/[W_{n2}V_{n1}+W_{n1}V_{n2}]. \tag{10-6b}$$

In these equations, as well as for the remainder of this section, the following functions are defined as

$$A_{n\tau} = \zeta J_n(\gamma_\tau a) [\partial J_n(k_o a)] - [\partial J_n(\gamma_\tau a)] J_n(k_o a); \quad \tau = 1,2, \tag{10-7a}$$

$$B_{n\tau} = \zeta [\partial J_n(\gamma_\tau a)] J_n(k_o a) - J_n(\gamma_\tau a) [\partial J_n(k_o a)]; \quad \tau = 1,2, \tag{10-7b}$$

$$V_{n\tau} = \zeta \, [\partial J_n(\gamma_\tau a)] \, H_n^{(1)}(k_o a) - J_n(\gamma_\tau a) \, [\partial H_n^{(1)}(k_o a)]; \quad \tau = 1,2, \tag{10-7c}$$

$$W_{n\tau} = \zeta \, J_n(\gamma_\tau a) \, [\partial H_n^{(1)}(k_o a)] - [\partial J_n(\gamma_\tau a)] \, H_n^{(1)}(k_o a); \quad \tau = 1,2, \tag{10-7d}$$

with $\zeta = \mu_o k / \mu k_o$ and $\partial J_n(\xi) = dJ_n(\xi)/d\xi$, etc. The extinction and the scattering efficiencies [van de Hulst 1981] for the present problem, respectively, are given as

$$Q_{ext} = (2/k_o a) \, Re \sum_{n \in [-\infty,\infty]} b_n, \tag{10-8a}$$

$$Q_{sca} = (2/k_o a) \sum_{n \in [-\infty,\infty]} |b_n|^2 + |c_n|^2. \tag{10-8b}$$

In the limit of no chirality, i.e., when $\beta = 0$, then $c_n = 0$ and the solution for nonchiral cylinder is recovered [Kerker 1969]

Exactly a similar procedure is to be followed for the TM case when the incident magnetic field is polarized parallel to the z axis, i.e., $E_{inc} = e_y \exp[ik_o x]$, and only the points of departure from the previous case are given. The incident electromagnetic field can be expanded as

$$E_{inc} = i \sum_{n \in [-\infty,\infty]} (i^n/k_o) \, M_n^{(1)}(k_o r) = e_y \exp[ik_o x], \tag{10-9a}$$

$$H_{inc} = (k_o/\omega\mu_o) \sum_{n \in [-\infty,\infty]} (i^n/k_o) \, N_n^{(1)}(k_o r), \tag{10-9b}$$

while the scattered field expansions are given by

$$E_{sc} = - \sum_{n \in [-\infty,\infty]} (i^n/k_o)\{ia_n M_n^{(3)}(k_o r) + d_n N_n^{(3)}(k_o r)\}, \tag{10-10a}$$

$$H_{sc} = (k_o/i\omega\mu_o) \sum_{n \in [-\infty,\infty]} (i^n/k_o)\{ia_n N_n^{(3)}(k_o r) + d_n M_n^{(3)}(k_o r)\}. \tag{10-10b}$$

The satisfaction of the boundary conditions leads to the solution

$$a_n = a_{-n} = [V_{n2}A_{n1} + V_{n1}A_{n2}]/[W_{n2}V_{n1} + W_{n1}V_{n2}], \tag{10-11a}$$

$$d_n = d_{-n} = i[W_{n1}A_{n2} - A_{n1}W_{n2}]/[W_{n2}V_{n1} + W_{n1}V_{n2}], \tag{10-11b}$$

while the extinction and the scattering efficiencies for this case, respectively, are given as

$$Q_{ext} = (2/k_o a) \, Re \sum_{n \in [-\infty,\infty]} a_n, \tag{10-12a}$$

45

$$Q_{sca} = (2/k_o a) \sum_{n \in [-\infty, \infty]} |a_n|^2 + |d_n|^2. \qquad (10\text{-}12b)$$

Again, in the limit of vanishing β, the coefficients d_n also vanish, while a_n is equivalent to that given by Kerker [1969] for non-chiral cylinders.

It can be readily shown that $c_n = -d_n$, so that there are only three independent scattering coefficient sets (viz., a_n, b_n and c_n) for the chiral cylinder, as will now be shown for chiral spheres as well.

11. SCATTERING BY A CHIRAL SPHERE

The scattering of electromagnetic waves by three-dimensional optically active objects is a quintessential problem in the biochemistry of chiral molecules [Urry & Krivacic 1970]. Before the advent of Bohren's decomposition, crude approximations for scattering by chiral spheres using the regular Mie theory for achiral spheres had been tried [Gordon & Holzwarth 1971; Gordon 1972; Holzwarth et al 1974] to varying degrees of success. But it was only after Bohren's pioneering work [Bohren 1974] that an exact solution for scattering by chiral spheres became available. And, as will be discussed in the next section, this work was extended by Lakhtakia et al [1985] using the T-matrix method [Waterman 1969, 1971] for any three-dimensional object.

Let a chiral sphere of radius a, embedded in free space, be illuminated by a plane wave, $E_{inc} = e_x \exp[ik_o z]$, as illustrated in Fig. 11-1. In terms of the spherical vector wave functions [Stratton 1941; Morse & Feshbach 1953], the incident field can be represented by

$$E_{inc} = \sum_{n \in [1,\infty]} \Lambda_n [M_{o1n}^{(1)}(k_o r) - iN_{e1n}^{(1)}(k_o r)], \qquad (11\text{-}1a)$$

$$H_{inc} = (-k_o/\omega\mu_o) \sum_{n \in [1,\infty]} \Lambda_n [M_{e1n}^{(1)}(k_o r) + iN_{o1n}^{(1)}(k_o r)], \qquad (11\text{-}1a)$$

in which the coefficient $\Lambda_n = i^n (2n+1)/n(n+1)$. The vector wave functions are defined as follows:

$$M_{e1n}^{(1)}(\sigma r) = \nabla \times [r \cos\varphi \, P_n^1(\cos\theta) \, j_n(\sigma r)], \qquad (11\text{-}2a)$$

$$M_{o1n}^{(1)}(\sigma r) = \nabla \times [r \sin\varphi \, P_n^1(\cos\theta) \, j_n(\sigma r)], \qquad (11\text{-}2b)$$

$$N_{e1n}^{(1)}(\sigma r) = (1/\sigma)\nabla \times M_{e1n}^{(1)}(\sigma r), \qquad (11\text{-}2c)$$

$$N_{o1n}^{(1)}(\sigma r) = (1/\sigma)\nabla \times M_{o1n}^{(1)}(\sigma r), \qquad (11\text{-}2d)$$

with $j_n(\sigma r)$ being the spherical Bessel function of order n, $P_n^1(\cos\theta)$ is the associated Legendre function of degree 1 and order n, $r = x e_x + y e_y + z e_z$, $\theta = \arctan(z/\sqrt{[x^2+y^2]})$ and $\varphi = \arctan(y/x)$.

Unlike in the case of an achiral sphere, the scattered field does not conserve the azimuthal parity when the sphere is chiral. Hence, the scattered field is expanded as

$$\mathbf{E}_{sc} = \sum_{n \in [1,\infty]} \Lambda_n \left[ia_n \mathbf{N}_{e1n}^{(3)}(k_o r) - b_n \mathbf{M}_{o1n}^{(3)}(k_o r) \right.$$
$$\left. + c_n \mathbf{M}_{e1n}^{(3)}(k_o r) - id_n \mathbf{N}_{o1n}^{(3)}(k_o r) \right], \tag{11-3a}$$

$$\mathbf{H}_{sc} = (k_o/\omega\mu_o) \sum_{n \in [1,\infty]} \Lambda_n \left[a_n \mathbf{M}_{e1n}^{(3)}(k_o r) + ib_n \mathbf{N}_{o1n}^{(3)}(k_o r) \right.$$
$$\left. - ic_n \mathbf{N}_{e1n}^{(3)}(k_o r) - d_n \mathbf{M}_{o1n}^{(3)}(k_o r) \right]. \tag{11-3b}$$

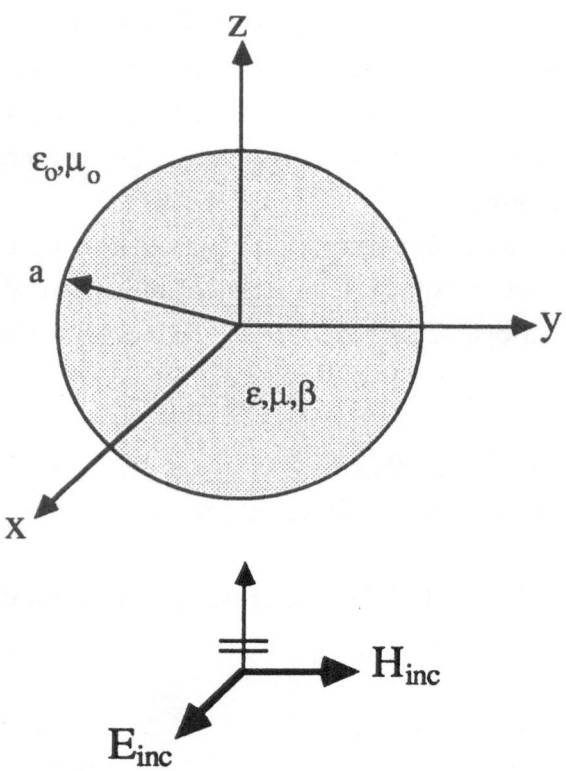

Fig. 11-1. Relevant to the scattering of a plane wave by a chiral sphere.

These expansions are valid on and outside the sphere $r = a;$, and the coefficients a_n, b_n, c_n and d_n are to be determined as a solution of the boundary value problem. The vector wave functions superscripted [3] in (11-3) can be obtained from the definitions (11-2) by replacing $j_n(\sigma r)$ by $h_n^{(1)}(\sigma r)$, the spherical Hankel functions of the first kind.

Finally, *vide* Bohren's decomposition, the field inside the chiral sphere is given by

$$Q_1 = \Sigma_{n \in [1,\infty]} \Lambda_n [f_{on}\{M_{o1n}^{(1)}(\gamma_1 r) + N_{o1n}^{(1)}(\gamma_1 r)\}$$
$$+ f_{en}\{M_{e1n}^{(1)}(\gamma_1 r) + N_{e1n}^{(1)}(\gamma_1 r)\}]; \qquad r \le a, \qquad (11\text{-}4a)$$

$$Q_2 = \Sigma_{n \in [1,\infty]} \Lambda_n [g_{on}\{M_{o1n}^{(1)}(\gamma_2 r) - N_{o1n}^{(1)}(\gamma_2 r)\}$$
$$+ g_{en}\{M_{e1n}^{(1)}(\gamma_2 r) - N_{e1n}^{(1)}(\gamma_2 r)\}]; \qquad r \le a, \qquad (11\text{-}4b)$$

with f_{on}, f_{en}, g_{on} and g_{en} being the unknown coefficients of expansion.

The solution is obtained by ensuring that there are no discontinuities in the tangential **E** and **H** fields across the spherical surface $r = a$. The resulting solution is given by

$$a_n = [V_{n2}A_{n1}+V_{n1}A_{n2}]/[W_{n2}V_{n1}+W_{n1}V_{n2}], \qquad (11\text{-}5a)$$
$$b_n = [W_{n2}B_{n1}+W_{n1}B_{n2}]/[W_{n2}V_{n1}+W_{n1}V_{n2}], \qquad (11\text{-}5b)$$
$$c_n = - d_n = i[W_{n2}A_{n1}-W_{n1}A_{n2}]/[W_{n2}V_{n1}+W_{n1}V_{n2}]. \qquad (11\text{-}5c)$$

In the foregoing expressions, the following functions have ben used:

$$A_{n\tau} = \zeta \, \psi_n(\gamma_\tau a) [\partial \psi_n(k_o a)] - [\partial \psi_n(\gamma_\tau a)] \, \psi_n(k_o a); \quad \tau = 1,2, \qquad (11\text{-}6a)$$
$$B_{n\tau} = \psi_n(\gamma_\tau a) [\partial \psi_n(k_o a)] - \zeta [\partial \psi_n(\gamma_\tau a)] \, \psi_n(k_o a); \quad \tau = 1,2, \qquad (11\text{-}6b)$$
$$V_{n\tau} = \psi_n(\gamma_\tau a) [\partial \xi_n(k_o a)] - \zeta [\partial \psi_n(\gamma_\tau a)] \, \xi_n(k_o a); \quad \tau = 1,2, \qquad (11\text{-}6c)$$
$$W_{n\tau} = \zeta \, \psi_n(\gamma_\tau a) [\partial \xi_n(k_o a)] - [\partial \psi_n(\gamma_\tau a)] \, \xi_n(k_o a); \quad \tau = 1,2, \qquad (11\text{-}6d)$$

with $\zeta = \mu_o k/\mu k_o$ and $\partial \psi_n(\alpha) = d\psi_n(\alpha)/d\alpha$, etc.; the Riccati-Bessel functions $\psi_n(\alpha) = \alpha \, j_n(\alpha)$ and $\xi_n(\alpha) = \alpha \, h_n^{(1)}(\alpha)$. In the limit of no chirality, i.e., when $\beta = 0$, then $c_n = d_n = 0$, and the solution reverts to the Mie solution for achiral spheres. It needs be mentioned that there are only three independent scattering coefficient sets (viz., a_n, b_n and c_n) for the chiral sphere, as was also the case for chiral cylinders of the previous section.

Using the forward scattering amplitude theorem [Jackson 1975], the extinction efficiency works out as

$$Q_{ext} = (2/k_o^2 a^2) \, Re \sum_{n \in [-\infty, \infty]} (2n+1)(a_n + b_n), \qquad (11\text{-}7a)$$

while the scattering efficiency has been shown to be

$$Q_{sca} = (2/k_o^2 a^2) \sum_{n \in [-\infty, \infty]} (2n+1)(|a_n|^2 + |b_n|^2 + |c_n|^2 + |d_n|^2). \qquad (11\text{-}7b)$$

The solution of this problem is due to Bohren [1974, 1975a], who has also studied the scattering response of spherical chiral shells [Bohren 1975b] using the presented formalism. In the next section, the T-matrix method [Waterman 1971] has been used to consider scattering by a chiral object of arbitrary shape embedded in free space.

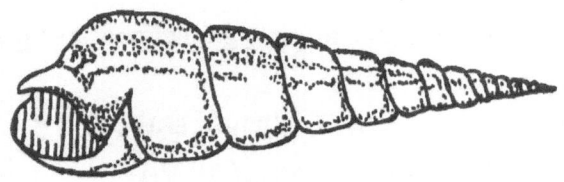

12. SCATTERING BY 3-D CHIRAL BODIES - THE T-MATRIX METHOD

The overall goal in using the T-matrix procedure is to determine the fields induced inside a permeable scatterer of arbitrary shape, as well as the fields scattered by it, when it is illuminated by an incident electromagentic field. Though the first account of the application of this method for an achiral dielectric object was given by Waterman [1969, 1971], a conceptually simpler approach due to Barber and Yeh [1975] has been utilized by Lakhtakia et al [1985] to investigate scattering by a three-dimensional chiral scatterer of arbitrary shape shown in Fig. 12-1.

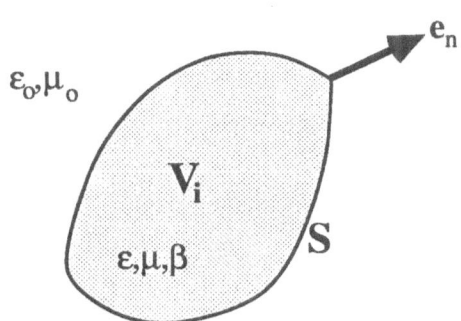

Fig. 12-1. Relevant to the T-matrix method for scattering by 3-D chiral bodies.

The total field existing outside the volume V_i of the scatterer can be expressed as

$$E(r) = E_{inc}(r) + E_{sc}(r); \quad r \notin V_i, \tag{12-1}$$

where the field E_{sc} is the scattered field and E_{inc} is the incident field. The T-matrix equations are then given as

$$E_{sc}(r) = \nabla \times \int_S ds' \, e_n \times E(r') \cdot \mathbb{G}(k_o|r,r')$$
$$- \nabla \times \nabla \times \int_S ds'(1/i\omega\varepsilon_o) \, e_n \times H(r') \cdot \mathbb{G}(k_o|r,r') \, ds'; \, r \notin V_i \tag{12-2}$$

$$E_{inc}(r) = - \nabla \times \int_S ds' \, e_n \times E(r') \cdot \mathfrak{G}(k_o|r,r')$$
$$+ \nabla \times \nabla \times \int_S ds'(1/i\omega\epsilon_o) \, e_n \times H(r') \cdot \mathfrak{G}(k_o|r,r') \, ds'; \, r \in V_i \quad (12\text{-}3)$$

where $\mathfrak{G}(k_o|r,r')$ is the free space Green's dyadic [Morse & Feshbach 1953], S is the scatterer surface and e_n is the unit outward normal to S, In particular, (12-3) is commonly called the *extinction theorem* and is the bedrock of the T-matrix method.

The usual boundary conditions must be satisfied by the solution. Therefore, if E_{int} and H_{int} are the fields induced inside the chiral scatterer, then (12-2) and (12-3) can be restated, respectively, as

$$E_{sc}(r) = \nabla \times \int_S ds' \, e_n \times E_{int}(r') \cdot \mathfrak{G}(k_o|r,r')$$
$$- \nabla \times \nabla \times \int_S ds'(1/i\omega\epsilon_o) \, e_n \times H_{int}(r') \cdot \mathfrak{G}(k_o|r,r') \, ds'; \, r \notin V_i, \quad (12\text{-}4)$$

$$E_{inc}(r) = - \nabla \times \int_S ds' \, e_n \times E_{int}(r') \cdot \mathfrak{G}(k_o|r,r')$$
$$+ \nabla \times \nabla \times \int_S ds'(1/i\omega\epsilon_o) \, e_n \times H_{int}(r') \cdot \mathfrak{G}(k_o|r,r') \, ds'; \, r \in V_i. \quad (12\text{-}5)$$

These two equations must be simultaneously solved to yield the internally induced and the scattered fields.

The incident field can be purely arbitrary, except that it should not have any singularities inside a minimum sphere circumscribing the scatterer; it is expanded in terms of vector spherical wave functions $M_v^{(1)}(k_o r)$ and $N_v^{(1)}(k_o r)$ [Morse & Feshbach 1953], regular at the origin as

$$E_{inc}(r) = \sum\nolimits_{v=smn} D_{nm}[a_v \, M_v^{(1)}(k_o r) + b_v \, N_v^{(1)}(k_o r)], \quad (12\text{-}6a)$$
$$H_{inc}(r) = (1/i\omega\mu_o) \, \nabla \times E_{inc}(r), \quad (12\text{-}6b)$$

where $v = smn$ is a triple-index. The index s can be even or odd, the index n varies from 1 to ∞ and the index m assumes values from 0 to n; a_v and b_v are the known expansion coefficients of the incident field, while D_{nm} is a normalization constant given by

$$D_{nm} = (2 - \delta_{m0}) \, \{(2n+1) / [4n(n+1)]\} \, \{(n-m)! / (n+m)!\}, \quad (12\text{-}7)$$

52

$\delta_{mm'}$ being the Kronecker delta. The vector wave functions used in (12-6) are defined as

$$\mathbf{M}_{emn}^{(1)}(\sigma r) = - (m/\sin\theta)\, j_n(\sigma r)\, P_n^m(\cos\theta)\, \sin m\varphi\; \mathbf{e}_\theta$$
$$- j_n(\sigma r)\, \partial P_n^m(\cos\theta)\, \cos m\varphi\; \mathbf{e}_\varphi, \qquad (12\text{-}8a)$$

$$\mathbf{M}_{omn}^{(1)}(\sigma r) = (m/\sin\theta)\, j_n(\sigma r)\, P_n^m(\cos\theta)\, \cos m\varphi\; \mathbf{e}_\theta$$
$$- j_n(\sigma r)\, \partial P_n^m(\cos\theta)\, \sin m\varphi\; \mathbf{e}_\varphi, \qquad (12\text{-}8b)$$

$$\mathbf{N}_{emn}^{(1)}(\sigma r) = (1/\sigma)\nabla\times\mathbf{M}_{emn}^{(1)}(\sigma r), \qquad (12\text{-}8c)$$

$$\mathbf{N}_{omn}^{(1)}(\sigma r) = (1/\sigma)\nabla\times\mathbf{M}_{omn}^{(1)}(\sigma r), \qquad (12\text{-}8d)$$

with $j_n(\sigma r)$ being the spherical Bessel function of order n and $P_n^m(\cos\theta)$ being the associated Legendre function of degree m and order n; $\partial P_n^m = dP_n^m/d\theta$.

Likewise, the scattered fields are expanded in terms of the functions $\mathbf{M}_v^{(3)}(k_o r)$ and $\mathbf{N}_v^{(3)}(k_o r)$ which obey Sommerfeld's radiation condition at infinity. Thus,

$$\mathbf{E}_{sc}(\mathbf{r}) = \sum_{v=smn} D_{nm}[c_v\, \mathbf{M}_v^{(3)}(k_o r) + d_v\, \mathbf{N}_v^{(4)}(k_o r)], \qquad (12\text{-}9a)$$

$$\mathbf{H}_{sc}(\mathbf{r}) = (1/i\omega\mu_o)\, \nabla\times \mathbf{E}_{sc}(\mathbf{r}), \qquad (12\text{-}9b)$$

this expansion being valid outside a sphere circumscribing the scatterer, and c_v and d_v being the unknown coefficients to be determined. Finally, the free space dyadic Green's function is given by

$$\mathbf{\Phi}(k_o|\mathbf{r},\mathbf{r'}) = (ik_o/\pi)\sum_{v=smn} D_{nm}\big\{ \mathbf{M}_v^{(3)}(k_o r_>)\, \mathbf{M}_v^{(1)}(k_o r_<)$$
$$+ \mathbf{N}_v^{(3)}(k_o r_>)\, \mathbf{N}_v^{(1)}(k_o r_<)\big\}, \qquad (12\text{-}10)$$

where $\mathbf{r}_>$ and $\mathbf{r}_<$, respectively, are the greater and the lesser of $\mathbf{r}$ and $\mathbf{r'}$ in magnitude. The vector wave functions superscripted (3) in (12-9) and (12-10) can be obtained from the definitions (12-8) by replacing $j_n(\sigma r)$ by $h_n^{(1)}(\sigma r)$, the spherical Hankel functions of the first kind.

Substitution of (12-6), (12-9) and (12-10) into (12-4) and (12-5), after applying some analytic continuation arguments [Waterman 1971], leads to:

53

$$a_v = (-ik_o^2/\pi) \left\{ \int_S ds' \, [e_n \times E_{int}(r') \cdot N_v^{(3)}(k_or') - (k_o/i\omega\varepsilon_o) \, e_n \times H_{int}(r') \cdot M_v^{(3)}(k_or')] \right\},$$ (12-11a)

$$b_v = (-ik_o^2/\pi) \left\{ \int_S ds' \, [e_n \times E_{int}(r') \cdot M_v^{(3)}(k_or') - (k_o/i\omega\varepsilon_o) \, e_n \times H_{int}(r') \cdot N_v^{(3)}(k_or')] \right\},$$ (12-11b)

$$c_v = (ik_o^2/\pi) \left\{ \int_S ds' \, [e_n \times E_{int}(r') \cdot N_v^{(1)}(k_or') - (k_o/i\omega\varepsilon_o) \, e_n \times H_{int}(r') \cdot M_v^{(1)}(k_or')] \right\},$$ (12-11c)

$$d_v = (ik_o^2/\pi) \left\{ \int_S ds' \, [e_n \times E_{int}(r') \cdot M_v^{(1)}(k_or') - (k_o/i\omega\varepsilon_o) \, e_n \times H_{int}(r') \cdot N_v^{(1)}(k_or')] \right\}.$$ (12-11d)

The remaining part of this problem is now to find adequate representations of the functions Q_1 and Q_2 which satisfy Bohren's requirements; these, again, are expansions in vector spherical harmonics:

$$Q_1(r) = \sum_{v=smn} f_v \, [M_v^{(1)}(\gamma_1 r) + N_v^{(1)}(\gamma_1 r)],$$ (12-12a)
$$Q_2(r) = \sum_{v=smn} g_v \, [M_v^{(1)}(\gamma_2 r) - N_v^{(1)}(\gamma_2 r)],$$ (12-12b)

with f_v and g_v being as yet unknown. Substitution of (7-1) and (12-12) into (12-11) yields, in matrix notation,

$$[a; b]^{(tr)} = [Y_3] \, [f; g]^{(tr)},$$ (12-13a)
$$[c; d]^{(tr)} = [Y_1] \, [f; g]^{(tr)},$$ (12-13b)

where the superscripted qualifier $^{(tr)}$ denotes the transpose, and the matrices Y_1 and Y_3 are of the forms

$$[Y_1] = - \begin{bmatrix} I^{(1)} & J^{(1)} \\ K^{(1)} & L^{(1)} \end{bmatrix} \quad ; \quad [Y_3] = \begin{bmatrix} I^{(3)} & J^{(3)} \\ K^{(3)} & L^{(3)} \end{bmatrix},$$ (12-14)

The elements of the submatrices $I^{(3)}$, etc., are integrals performed over the surface

S of the chiral scatterer and, after some manipulation, simplify to:

$$I^{(3)}_{vv'} = (ik_o{}^2/\pi) \left\{ \int_S ds' \, e_n \times [M_{v'}^{(1)}(\gamma_1 r') + N_{v'}^{(1)}(\gamma_1 r')] \right.$$
$$\left. \cdot [N_v^{(3)}(k_o r') + \xi M_v^{(3)}(k_o r')] \right\}, \qquad (12\text{-}15a)$$

$$J^{(3)}_{vv'} = (ik_o{}^2/\pi) \left\{ \int_S ds' \, e_n \times [M_{v'}^{(1)}(\gamma_2 r') - N_{v'}^{(1)}(\gamma_1 r')] \right.$$
$$\left. \cdot [N_v^{(3)}(k_o r') - \xi M_v^{(3)}(k_o r')] \right\} a_R, \quad (12\text{-}15b)$$

$$K^{(3)}_{vv'} = (ik_o{}^2/\pi) \left\{ \int_S ds' \, e_n \times [M_{v'}^{(1)}(\gamma_1 r') + N_{v'}^{(1)}(\gamma_1 r')] \right.$$
$$\left. \cdot [M_v^{(3)}(k_o r') + \xi N_v^{(3)}(k_o r')] \right\}, \qquad (12\text{-}15c)$$

$$L^{(3)}_{vv'} = (ik_o{}^2/\pi) \left\{ \int_S ds' \, e_n \times [M_{v'}^{(1)}(\gamma_2 r') - N_{v'}^{(1)}(\gamma_1 r')] \right.$$
$$\left. \cdot [M_v^{(3)}(k_o r') - \xi N_v^{(3)}(k_o r')] \right\} a_R, \quad (12\text{-}15d)$$

where $\xi = [(\mu_o/\varepsilon_o)(\varepsilon/\mu)]^{1/2}$ and a_R is defined by (7-2). In order to obtain the submatrices $I^{(1)}$, etc., the superscripted qualifier $^{(3)}$ in (12-15) should be replaced by $^{(1)}$. Finally, the scattered field coefficients can be related to those of the incident field *via* a T-matrix:

$$[c; d]^{(tr)} = [Y_1][Y_3]^{-1} [a; b]^{(tr)} = [T] [a; b]^{(tr)}, \qquad (12\text{-}16)$$

provided the inverse of Y_3 can be computed.

Provided that the scatterer is rotationally symmetric about the z axis of the chosen coordinate system, the matrix equations (12-13) simplify considerably due to the trivialization of the integrals on the polar variable φ. In particular, the integrals (12-15) contain factors of the type $\delta_{mm'}$, which indicates that the matrix equations (12-13) decouple in the azimuthal index m. This allows the solution of (12-13) for each value of m separately. Furthermore, these integrals also contain factors of the type $\delta_{ss'}$ and $(1 - \delta_{ss'})$ so that further reductions due to even-even, even-odd, odd-even, and odd-odd parities also take place. Consideration of these reductions leads to the transformation of (12-13a) to

$$
\begin{bmatrix} a_{emn} \\ a_{omn} \\ b_{emn} \\ b_{omn} \end{bmatrix} = \begin{bmatrix} A^{(3)}_{mnn'}(\gamma_1) & -B^{(3)}_{mnn'}(\gamma_1) & A^{(3)}_{mnn'}(\gamma_2) & B^{(3)}_{mnn'}(\gamma_2) \\ B^{(3)}_{mnn'}(\gamma_1) & A^{(3)}_{mnn'}(\gamma_1) & -B^{(3)}_{mnn'}(\gamma_2) & A^{(3)}_{mnn'}(\gamma_2) \\ C^{(3)}_{mnn'}(\gamma_1) & -D^{(3)}_{mnn'}(\gamma_1) & -C^{(3)}_{mnn'}(\gamma_2) & -D^{(3)}_{mnn'}(\gamma_2) \\ D^{(3)}_{mnn'}(\gamma_1) & C^{(3)}_{mnn'}(\gamma_1) & D^{(3)}_{mnn'}(\gamma_2) & -C^{(3)}_{mnn'}(\gamma_2) \end{bmatrix} \begin{bmatrix} f_{emn'} \\ f_{omn'} \\ g_{emn'} a_R \\ g_{omn'} a_R \end{bmatrix} \quad (12\text{-}17)
$$

where

$$
A^{(3)}_{mnn'}(p) = (-ik_o^2/\pi) \Big\{ \int_S ds' e_n \bullet ([M_{emn'}^{(1)}(pr') \times N_{emn}^{(3)}(k_o r')]
$$
$$
+ [\xi N_{emn'}^{(1)}(pr') \times M_{emn}^{(3)}(k_o r')]) \Big\}, \quad (12\text{-}18a)
$$

$$
B^{(3)}_{mnn'}(p) = (-ik_o^2/\pi) \Big\{ \int_S ds' e_n \bullet ([N_{emn'}^{(1)}(pr') \times N_{omn}^{(3)}(k_o r')]
$$
$$
+ [\xi M_{emn'}^{(1)}(pr') \times M_{omn}^{(3)}(k_o r')]) \Big\}, \quad (12\text{-}18b)
$$

$$
C^{(3)}_{mnn'}(p) = (-ik_o^2/\pi) \Big\{ \int_S ds' e_n \bullet ([N_{emn'}^{(1)}(pr') \times M_{emn}^{(3)}(k_o r')]
$$
$$
+ [\xi M_{emn'}^{(1)}(pr') \times N_{emn}^{(3)}(k_o r')]) \Big\}, \quad (12\text{-}18c)
$$

$$
D^{(3)}_{mnn'}(p) = (-ik_o^2/\pi) \Big\{ \int_S ds' e_n \bullet ([M_{emn'}^{(1)}(pr') \times M_{omn}^{(3)}(k_o r')]
$$
$$
+ [\xi N_{emn'}^{(1)}(pr') \times N_{omn}^{(3)}(k_o r')]) \Big\}. \quad (12\text{-}18d)
$$

The corresponding analog of (12-13b) can be easily obtained by replacing the vector $[a_{emn}; a_{omn}; b_{emn}; b_{omn}]^{(tr)}$ in (12-17) by $[-c_{emn}; -c_{omn}; -d_{emn}; -d_{omn}]^{(tr)}$, and by replacing the superscripted qualifier $^{(3)}$ everywhere in (12-17) and (12-18) by the qualifier $^{(1)}$. These reductions result in a great deal of computational economy; hence, their utility.

It must be noted that in case the scatterer is spherical, the submatrices $A^{(3)}_{mnn'}$, etc. become diagonal; i.e., they involve $\delta_{nn'}$ and $(1 - \delta_{nn'})$ as factors. Consequently, the matrix T of (12-16) also becomes diagonal, and resulting solution tallies exactly with that for chiral spheres. In addition, should $\beta = 0$ also hold, i.e., the scatterer becomes a non-chiral sphere, the present procedure degenerates into the well-known Mie series solution. Lastly, simply by setting $\beta = 0$, solutions for non-chiral, nonspherical objects are also recoverable from the presented technique.

The total scattering efficiency [van de Hulst 1981] can be computed by integrating the scattered Poynting vector in the far zone over the whole solid angle, and is given by

$$Q_{sca} = (k_o a)^{-2} \sum_{v=snm} D_{nm} \{|c_v|^2 + |d_v|^2\}, \tag{12-19}$$

while the absorption efficiency has to be evaluated by the surface integral

$$Q_{abs} = (120/a^2) \, Re \left\{ \int_S ds \, e_n \cdot E_{int}(r) \times H_{int}^*(r) \right\}, \tag{12-20}$$

so that the extinction efficiency can be obtained *via*

$$Q_{ext} = Q_{sca} + Q_{abs}. \tag{12-21}$$

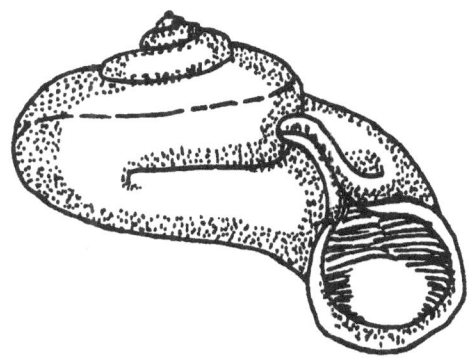

13. INFINITE-MEDIUM DYADIC GREEN'S FUNCTIONS FOR THE ELECTROMAGNETIC FIELDS

Bohren's decomposition is extremely useful for investigating scattering by bounded chiral volumes, but a systematic study of electromagnetic wave propagation characteristics in chiral media requires knowledge of the *infinite medium* Green's function, which is, of course, nothing but a field propagator. In order to find a *dyadic* Green's function for chiral media, utilization can be made of the dyadic formalism [Chen 1983], which has also been employed for the same purpose by Bassiri et al [1986] for the isotropic Post constitutive equations (3-8).

To begin with, recall (4-10), which are restated here for convenience:

$$\nabla\times\nabla\times E - 2\gamma^2\beta \ \nabla\times E - \gamma^2 E = i\omega\mu(\gamma/k)^2[J + \beta\nabla\times J] - (\gamma/k)^2\nabla\times K, \quad (4\text{-}10a)$$
$$\nabla\times\nabla\times H - 2\gamma^2\beta\nabla\times H - \gamma^2 H = i\omega\epsilon(\gamma/k)^2[K + \beta\nabla\times K] + (\gamma/k)^2\nabla\times J; \quad (4\text{-}10b)$$

thus, dyadic Green's functions are required for two equations:

$$[\nabla\nabla - \nabla^2 \mathbf{I} - \gamma^2 \mathbf{I} - 2\gamma^2\beta \ \nabla\times\mathbf{I}]\cdot U(r) = U_o(r), \qquad (13\text{-}1)$$
$$[\nabla\nabla - \nabla^2 \mathbf{I} - \gamma^2 \mathbf{I} - 2\gamma^2\beta \ \nabla\times\mathbf{I}]\cdot U(r) = \nabla\times U_o(r), \qquad (13\text{-}2)$$

U_o being a source term, and $\mathbf{I}$ is the unit dyadic. Since (13-1) is linear, it must admit a solution of the form

$$U(r) \ = \int_V dv' \ \textcircled{B}(r,r') \cdot U_o(r'), \qquad (13\text{-}3)$$

with $\textcircled{B}$ being the *infinite medium* Green's dyadic, V the volume over which the source is distributed, **r** the field point and **r'** the source point. But, with $\delta(r - r')$ being the three-dimensional Dirac delta function,

$$U_o(r) = \int_V dv' \ U_o(r') \ \delta(r - r'), \qquad (13\text{-}4)$$

wherefore the substitution of (13-3) and (13-4) into (13-1) leads to the differential relation

58

$$\left[\nabla\nabla - \nabla^2\mathbf{I} - \gamma^2\mathbf{I} - 2\gamma^2\beta\ \nabla\times\mathbf{I}\right]\cdot\mathbf{G}(\mathbf{r},\mathbf{r}') = \mathbf{I}\delta(\mathbf{r} - \mathbf{r}'). \qquad (13\text{-}5)$$

Substituting, next, the three-dimensional Fourier transform

$$\mathbf{G}(\mathbf{r},\mathbf{r}') = (2\pi)^{-3}\ _{-\infty}\!\!\iiint^{\infty} d^3p\ \mathbf{g}(\mathbf{p})\ \exp[i\mathbf{p}\cdot(\mathbf{r} - \mathbf{r}')], \qquad (13\text{-}6\text{a})$$
$$\delta(\mathbf{r} - \mathbf{r}') = (2\pi)^{-3}\ _{-\infty}\!\!\iiint^{\infty} d^3p\ \exp[i\mathbf{p}\cdot(\mathbf{r} - \mathbf{r}')], \qquad (13\text{-}6\text{b})$$

in (13-5), the (algebraic) dyadic relation

$$\left[(\gamma^2 - p^2)\ \mathbf{I} + \mathbf{pp} + 2i\gamma^2\beta\ \mathbf{p}\times\mathbf{I}\right]\cdot\ \mathbf{g}(\mathbf{p}) = -\mathbf{I}. \qquad (13\text{-}7)$$

is obtained.

Now, the inverse of the dyadic $\mathbf{W} = \lambda\mathbf{I} + \mathbf{ab} + \mathbf{c}\times\mathbf{I}$ is given by $\mathbf{W}^{-1} = (\mathrm{adj}\,\mathbf{W})$ $[|\mathbf{W}|]^{-1}$, where

$$\mathrm{adj}\,\mathbf{W} = \lambda^2\mathbf{I}\ - \lambda[\mathbf{ab} - (\mathbf{a}\cdot\mathbf{b})\mathbf{I} + \mathbf{c}\times\mathbf{I}] + \mathbf{cc} - (\mathbf{b}\cdot\mathbf{c})(\mathbf{a}\times\mathbf{I}) - \mathbf{c}(\mathbf{a}\times\mathbf{b}), \quad (13\text{-}8\text{a})$$
$$|\mathbf{W}| = \lambda^3 + (\mathbf{a}\cdot\mathbf{b})\lambda^2 + (\mathbf{c}\cdot\mathbf{c} - \mathbf{a}\times\mathbf{b}\cdot\mathbf{c})\lambda + (\mathbf{a}\cdot\mathbf{c})(\mathbf{b}\cdot\mathbf{c}). \qquad (13\text{-}8\text{b})$$

On employing these relations in (13-7), and after some algebraic tedium, the relation

$$\mathbf{g}(\mathbf{p}) = [(\gamma^2\text{-}p^2)^2 - 4\gamma^4\beta^2p^2]^{-1}$$
$$\left\{ (\gamma^2\text{-}p^2)[\ \gamma^2\mathbf{pp} - \mathbf{I}]+2\gamma^2\beta[i\mathbf{p}\times\mathbf{I} + 2\beta\mathbf{pp}]\right\}, \qquad (13\text{-}9)$$

is obtained. The quantity $\mathbf{g}(\mathbf{p})$, together with (13-6a) gives a Green's function in the three-dimensional spatial frequency space, and has been anticipated by Kong [1972] as well as by Pattanayak and Birman [1981a]. But (13-9) can be transformed back, *vide* (13-6a), to give

$$8\pi^3\ \mathbf{G}(\mathbf{r},\mathbf{r}') = -\ [\gamma^{-2}\ \nabla\nabla + \mathbf{I}]\ K_1 + 2\gamma^2\beta[\nabla\times\mathbf{I} - 2\beta\nabla\nabla]\ K_2. \qquad (13\text{-}10)$$

In (13-10), the integrals K_1 and K_2 are as follows:

$$K_1 = {\scriptstyle -\infty}\!\!\iiint^\infty d^3p \; (\gamma^2-p^2)\big[(\gamma^2-p^2)^2 - 4\gamma^4\beta^2p^2\big]^{-1} \; \exp[i\mathbf{p}\cdot\mathbf{R}]$$
$$= (2\pi/iR) {\scriptstyle -\infty}\!\!\int^\infty dp \; (\gamma^2-p^2)\big[(\gamma^2-p^2)^2 - 4\gamma^4\beta^2p^2\big]^{-1} \; p\,\exp[ipR], \quad (13\text{-}11a)$$

$$K_2 = {\scriptstyle -\infty}\!\!\iiint^\infty d^3p \; \big[(\gamma^2-p^2)^2 - 4\gamma^4\beta^2p^2\big]^{-1} \; \exp[i\mathbf{p}\cdot\mathbf{R}]$$
$$= (2\pi/iR) {\scriptstyle -\infty}\!\!\int^\infty dp \; \big[(\gamma^2-p^2)^2 - 4\gamma^4\beta^2p^2\big]^{-1} \; p\,\exp[ipR], \quad (13\text{-}11b)$$

with $\mathbf{R} = \mathbf{r} - \mathbf{r}'$. The evaluation of these integrals must be done in the upper half of the complex plane. To be noted is the fact that the integrands of both K_1 and K_2 contain singularities at $p = \pm\gamma_1, \pm\gamma_2$. Since fields cannot increase in amplitude with the passage of time, of these four singularities two must be excluded because of the chosen time-dependence, $\exp(-i\omega t)$; thus, using Cauchy's residue theorem to evaluate the integrals of (13-11) in the complex p-plane, we obtain

$$K_1 = (2\pi^2/R) \left\{ (\gamma^2 - \gamma_1{}^2)\, \exp[i\gamma_1 R] \right.$$
$$\left. - (\gamma^2 - \gamma_2{}^2)\, \exp[i\gamma_2 R] \right\} \; [\gamma_1{}^2 - \gamma_2{}^2]^{-1}, \quad (13\text{-}12a)$$
$$K_2 = (2\pi^2/R) \left\{ \exp[i\gamma_1 R] - \exp[i\gamma_2 R] \right\} \; [\gamma_1{}^2 - \gamma_2{}^2]^{-1}. \quad (13\text{-}12b)$$

These two integrals are substituted back into (13-10), whence it turns out that

$$\mathbf{\mathcal{G}}(\mathbf{r}, \mathbf{r}') = \mathbf{\mathcal{G}}_1(\mathbf{r}, \mathbf{r}') + \mathbf{\mathcal{G}}_2(\mathbf{r}, \mathbf{r}'), \quad (13\text{-}13a)$$
$$\mathbf{\mathcal{G}}_1(\mathbf{r}, \mathbf{r}') = (k/8\pi\gamma^2) \left[\gamma_1\mathbf{J} + \gamma_1^{-1} \nabla\nabla + \nabla\times\mathbf{J}\right] g(\gamma_1; R), \quad (13\text{-}13b)$$
$$\mathbf{\mathcal{G}}_2(\mathbf{r}, \mathbf{r}') = (k/8\pi\gamma^2) \left[\gamma_2\mathbf{J} + \gamma_2^{-1} \nabla\nabla - \nabla\times\mathbf{J}\right] g(\gamma_2; R), \quad (13\text{-}13c)$$
$$g(\sigma; R) = \exp[i\sigma R]/R. \quad (13\text{-}13d)$$

It must be mentioned that only values of β which cause both γ_1 and γ_2 to have finite, positive real parts are to be considered here. In the limiting cases when $k\beta = \pm 1$, the Green's dyadic is given by

$$\mathbf{\mathcal{G}}(\mathbf{r}, \mathbf{r}')|_{k\beta = \pm 1} =$$
$$-(1/8\pi k) \left[-(k/2)\mathbf{J} - (k/2)^{-1} \nabla\nabla \pm \nabla\times\mathbf{J}\right] g(k/2; R), \quad (13\text{-}14)$$

Two reciprocity properties of the dyadic Green's function should be noted, *viz.*,

$$\mathfrak{G}(\mathbf{r}, \mathbf{r}') = [\mathfrak{G}(\mathbf{r}', \mathbf{r})]^{(tr)}, \tag{13-15a}$$

$$\nabla \times \mathfrak{G}(\mathbf{r}, \mathbf{r}') = [\nabla' \times \mathfrak{G}(\mathbf{r}', \mathbf{r})]^{(tr)}, \tag{13-15b}$$

in which the superscripted qualifier $^{(tr)}$ denotes transpose. Furthermore,

$$\nabla \times \mathfrak{G}_1(\mathbf{r}, \mathbf{r}') = \gamma_1 \, \mathfrak{G}_1(\mathbf{r}, \mathbf{r}'), \tag{13-16a}$$

$$\nabla \times \mathfrak{G}_2(\mathbf{r}, \mathbf{r}') = -\gamma_2 \, \mathfrak{G}_2(\mathbf{r}, \mathbf{r}'), \tag{13-16b}$$

which show that the isotropic chiral media are circularly birefringent.

The rotational properties (13-16) can be utilized to yield yet another decomposition of the dyadic Green's function. By using these properties in (13-5) *a priori*, it can be shown that

$$\mathfrak{G}(\mathbf{r}, \mathbf{r}') = \Gamma_1(\mathbf{r}, \mathbf{r}') \oplus \Gamma_2(\mathbf{r}, \mathbf{r}'), \tag{13-17a}$$

$$\Gamma_1(\mathbf{r}, \mathbf{r}') = (1/4\pi) \left[\mathbf{I} + \gamma_1^{-2} \nabla\nabla \right] g(\gamma_1; R), \tag{13-17b}$$

$$\Gamma_2(\mathbf{r}, \mathbf{r}') = (1/4\pi) \left[\mathbf{I} + \gamma_2^{-2} \nabla\nabla \right] g(\gamma_2; R), \tag{13-17c}$$

in which the symbol $\oplus$ denotes an appropriate linear combination. The dyadics Γ_1 and Γ_2, respectively, should be used independently for the LCP and the RCP fields, $\mathbf{Q}_1$ and $\mathbf{Q}_2$, defined in Section 8. Along with this decomposition, however, the auxiliary conditions

$$\nabla \times \Gamma_1(\mathbf{r}, \mathbf{r}') = \gamma_1 \, \Gamma_1(\mathbf{r}, \mathbf{r}'), \tag{13-18a}$$

$$\nabla \times \Gamma_2(\mathbf{r}, \mathbf{r}') = -\gamma_2 \, \Gamma_2(\mathbf{r}, \mathbf{r}'), \tag{13-18b}$$

must be explicitly mentioned since they are not built into the definitions (13-17). Explicit use of the dyadics Γ_1 and Γ_2 will be made in Section 18.

Needless to say, should $\beta = 0$, then either from (13-13) or from (13-17) the Green's dyadic will have the familiar form,

$$\mathfrak{G}(\mathbf{r}, \mathbf{r}') = (1/4\pi) \left[\mathbf{I} + k^{-2} \nabla\nabla \right] g(k; R), \tag{13-19}$$

for isotropic, achiral media. It must be pointed out that the circular birefringence of the chiral media, *vide* (13-6), has been obtained here, not from examining the plane wave propagation as is commonly done, but from the Green's function: this implies that all kinds of fields -- near- as well as far-fields; fields with spherical, cylindrical, planar or any other wavefronts -- are circularly birefringent.

For numerical computational procedures like the spectral-domain iteration method [e.g., Kastener & Mittra 1983] and the method of moments [Harrington 1968], the dyadic Green's function can be exploited very easily by using the relations

$$\nabla\nabla g(\sigma;R) = \left[\{i\sigma R^{-1} - R^{-2}\}(\mathbf{J} - 3\,\mathbf{RR}\,R^{-2}) - (\sigma/R)^2\,\mathbf{RR} \right] g(\sigma;R), \quad (13\text{-}20)$$

$$\nabla\times[\mathbf{J}\,g(\sigma;R)] = \{i\sigma R^{-1} - R^{-2}\}\,g(\sigma;R)\,\mathbf{R}\times\mathbf{J}, \quad (13\text{-}21)$$

in the definitions (13-13).

For semi-analytical procedures like the T-matrix method [Waterman 1969, 1971; Varadan & Varadan 1980], the Green's function needs to be expanded in terms of global basis functions. In that case it turns out to be more convenient to use the decomposition (13-17) of the Green's dyadic, which yields

$$\Gamma_1(\mathbf{r}, \mathbf{r}') = (i\gamma_1/\pi) \sum_\nu D_{nm}\, \mathbf{L}_\nu^{(3)}(\mathbf{r}_>)\, \mathbf{L}_\nu^{(1)}(\mathbf{r}_<), \quad (13\text{-}22a)$$

$$\Gamma_2(\mathbf{r}, \mathbf{r}') = (i\gamma_2/\pi) \sum_\nu D_{nm}\, \mathbf{R}_\nu^{(3)}(\mathbf{r}_>)\, \mathbf{R}_\nu^{(1)}(\mathbf{r}_<), \quad (13\text{-}22b)$$

where $\mathbf{r}_>$ and $\mathbf{r}_<$, respectively, are the greater and the lesser of $\mathbf{r}$ and $\mathbf{r}'$ in magnitude. The functions $\mathbf{L}_\nu^{(j)}$ and $\mathbf{R}_\nu^{(j)}$ are defined in terms of the vector spherical harmonics $\mathbf{M}_\nu^{(j)}$ and $\mathbf{N}_\nu^{(j)}$, and are given by

$$\mathbf{L}_\nu^{(j)}(\mathbf{r}) = \mathbf{M}_\nu^{(j)}(\gamma_1 \mathbf{r}) + \mathbf{N}_\nu^{(j)}(\gamma_1 \mathbf{r}), \quad (13\text{-}23a)$$

$$\mathbf{R}_\nu^{(j)}(\mathbf{r}) = \mathbf{M}_\nu^{(j)}(\gamma_2 \mathbf{r}) - \mathbf{N}_\nu^{(j)}(\gamma_2 \mathbf{r}), \quad (13\text{-}23b)$$

where, $\mathbf{M}_\nu^{(j)}$ and $\mathbf{N}_\nu^{(j)}$ are the vector spherical wavefunctions defined in Section 12. It should be noted that $\nabla\times \mathbf{L}_\nu^{(j)} = \gamma_1 \mathbf{L}_\nu^{(j)}$ and $\nabla\times \mathbf{R}_\nu^{(j)} = -\gamma_2 \mathbf{R}_\nu^{(j)}$; hence the expressions (13-22) automatically satisfy the rotational properties (13-18). Justification for the expansions (13-22) has been obtained by Lakhtakia et al [1987b] by using the Helmholtz theorem on the sphere.

By direct substitution, it is a simple matter to verify that $\nabla \times \mathfrak{G}(\mathbf{r}, \mathbf{r}')$ is the appropriate Green's dyadic for (13-2), and will be frequently needed for the solution of radiation problems.

Expressions (13-13) define a three-dimensional Green's dyadic. If the electromagnetic problem were to be independent of the z co-ordinate, then $\partial/\partial z \equiv 0$. In that case, the *two-dimensional* Green's dyadic is given as

$$\mathfrak{G}(\mathbf{r}, \mathbf{r}') = (ik/8\gamma^2) \left[\gamma_1 \mathbf{I} + \gamma_1^{-1} \nabla\nabla + \nabla \times \mathbf{I} \right] H_0(\gamma_1 |\mathbf{r}-\mathbf{r}'|)$$
$$+ (ik/8\gamma^2) \left[\gamma_2 \mathbf{I} + \gamma_2^{-1} \nabla\nabla - \nabla \times \mathbf{I} \right] H_0 (\gamma_2 |\mathbf{r}-\mathbf{r}'|), \qquad (13\text{-}24)$$

where $H_0(\cdot)$ is the cylindrical Hankel function of the first kind and zero order, and $\mathbf{r}$ is the two-dimensional radial vector. Next, if both $\partial/\partial z \equiv 0$ and $\partial/\partial y \equiv 0$, then the *one-dimensional* Green's dyadic is

$$\mathfrak{G}(x, x') = (ik/4\gamma_1\gamma^2) \left[\gamma_1 \mathbf{I} + \gamma_1^{-1} \nabla\nabla + \nabla \times \mathbf{I} \right] \exp(i\gamma_1 |x - x'|)$$
$$+ (ik/4\gamma_2\gamma^2) \left[\gamma_2 \mathbf{I} + \gamma_2^{-1} \nabla\nabla - \nabla \times \mathbf{I} \right] \exp(i\gamma_2 |x - x'|). \qquad (13\text{-}25)$$

There is one last thing left to deal with in this section which has to do with the correctness of the expression (13-13a) of $\mathfrak{G}(\mathbf{r}, \mathbf{r}')$. The procedure utilized extracted the differential operators from a Fourier transform, as per (13-9) and (13-10), and lacks mathematical rigor. Hence, the need to prove that (13-13) indeed give the correct solution of (13-5).

It is well known that the scalar Green's function $g(\sigma;R)$ satisfies the differential equation

$$(\nabla^2 + \sigma^2) \, g(\sigma;R) = - 4\pi\delta(R), \qquad (13\text{-}26)$$

while the dyadic Green's function $\mathbf{g}(\sigma;R) = \{ \mathbf{I} + \nabla\nabla/\sigma^2 \} g(\sigma;R)$ satisfies the differential equation

$$[\nabla\nabla - \nabla^2 \mathbf{I} - \sigma^2 \mathbf{I}] \cdot \mathbf{g}(\sigma;R) = 4\pi \mathbf{I} \delta(R). \qquad (13\text{-}27)$$

By noting that divergence of the curl of any dyad as well as the curl of the gradient of any vector are identically zero, it becomes possible to deduce from (13-27) that

63

$$(\nabla^2 \mathbf{J} + \sigma^2 \mathbf{J})\, g(\sigma;R) = -\,4\pi \mathbf{J}\delta(\mathbf{R}); \tag{13-28a}$$

and by taking the curl of both sides of this latter equation, the following result is obtained:

$$[\nabla\nabla - \nabla^2\mathbf{J} - \sigma^2\mathbf{J}]\cdot[\nabla\times\mathbf{J}g(\sigma;R)] = 4\pi\nabla\times\mathbf{J}\delta(\mathbf{R}). \tag{13-28b}$$

After the foregoing preliminaries, and by direct use of (13-13b,c), the curl of $\mathbf{\mathfrak{G}}_1$ and $\mathbf{\mathfrak{G}}_2$ can be found as

$$\begin{aligned}
\nabla\times\mathbf{\mathfrak{G}}_1 = (k/8\pi\gamma^2)[\gamma\nabla\times\mathbf{J}g(\gamma_1;R) + \\
\nabla\times\nabla\nabla g(\gamma_1;R)/\gamma_1 + \nabla\times\nabla\times\mathbf{J}g(\gamma_1;R)], \tag{13-29a}
\end{aligned}$$
$$\begin{aligned}
\nabla\times\mathbf{\mathfrak{G}}_2 = (k/8\pi\gamma^2)[\gamma\nabla\times\mathbf{J}g(\gamma_2;R) + \\
\nabla\times\nabla\nabla g(\gamma_2;R)/\gamma_2 - \nabla\times\nabla\times\mathbf{J}g(\gamma_2;R)], \tag{13-29b}
\end{aligned}$$

whence, using (13-26) and the usual dyadic identities, it is easy to see that

$$\nabla\times\mathbf{\mathfrak{G}}_1 = \gamma_1\mathbf{\mathfrak{G}}_1 + (k/2\gamma^2)\mathbf{J}\delta(\mathbf{R}). \tag{13-30a}$$
$$\nabla\times\mathbf{\mathfrak{G}}_m = -\,\gamma_2\mathbf{\mathfrak{G}}_2 - (k/2\gamma^2)\mathbf{J}\delta(\mathbf{R}). \tag{13-30b}$$

With the help of the identities (13-30), one can evaluate that

$$\begin{aligned}
\left[\nabla\nabla - \nabla^2\mathbf{J} - \gamma^2\mathbf{J} - 2\gamma^2\beta\,\nabla\times\mathbf{J}\right]\cdot\mathbf{\mathfrak{G}}_1 \\
= [\,\nabla\nabla - \nabla^2\mathbf{J} - \gamma_2^2\,\mathbf{J}]\cdot\mathbf{\mathfrak{G}}_1 \;-\; k\beta\mathbf{J}\delta(\mathbf{R}), \tag{13-31a}
\end{aligned}$$
$$\begin{aligned}
\left[\nabla\nabla - \nabla^2\mathbf{J} - \gamma^2\mathbf{J} - 2\gamma^2\beta\,\nabla\times\mathbf{J}\right]\cdot\mathbf{\mathfrak{G}}_2 \\
= [\,\nabla\nabla - \nabla^2\mathbf{J} - \gamma_2^2\,\mathbf{J}]\cdot\mathbf{\mathfrak{G}}_2 \;+\; k\beta\mathbf{J}\delta(\mathbf{R}\}. \tag{13-31b}
\end{aligned}$$

Substitution of (13-13b,c) in (13-31a,b), respectively, followed by the use of (13-27) and (13-28b) in each case, leads to

$$\begin{aligned}
\left[\nabla\nabla - \nabla^2\mathbf{J} - \gamma^2\mathbf{J} - 2\gamma^2\beta\,\nabla\times\mathbf{J}\right]\cdot\mathbf{\mathfrak{G}}_1 \\
= (k\gamma_1/2\gamma^2)\mathbf{J}\delta(\mathbf{R}) + (k/2\gamma^2)[\nabla\times\mathbf{J}\delta(\mathbf{R})] - k\beta\mathbf{J}\delta(\mathbf{R}). \tag{13-32a}
\end{aligned}$$

$$\left[\nabla\nabla - \nabla^2 \mathbf{I} - \gamma^2 \mathbf{I} - 2\gamma^2\beta \ \nabla\times\mathbf{I}\right]\cdot\mathfrak{G}_2$$
$$= (k\gamma_2/2\gamma^2)\mathbf{I}\delta(\mathbf{R}) - (k/2\gamma^2)[\nabla\times\mathbf{I}\delta(\mathbf{R})] + k\beta\mathbf{I}\delta(\mathbf{R}). \qquad (13\text{-}32\text{b})$$

Thus, on using (13-13a), and (13-32), it turns out that

$$\left[\nabla\nabla - \nabla^2 \mathbf{I} - \gamma^2 \mathbf{I} - 2\gamma^2\beta \ \nabla\times\mathbf{I}\right]\cdot\mathfrak{G} = \mathbf{I}\delta(\mathbf{R}), \qquad (13\text{-}33)$$

which verifies the adequacy of the derived Green's function (13-13). Confirmation of (13-13) using operator algebra has also become available recently [Weiglhofer 1989]. In addition, a scalar treatment will be presented in Section 19, which will be shown to yield results identical to those derived using the vector-dyadic approach.

65

14. VECTOR AND SCALAR POTENTIALS

Several radiation problems for achiral media are considerably simplified by the use of scalar and vector potentials. In this section, the analogous potentials for isotropic chiral media will be introduced, followed by the derivation of the infinite-medium Green's function for these potentials.

The third Maxwell's equation, $\nabla \times E = i\omega B$, can be seen to be completely satisfied by a *vector magnetic potential* A and a *scalar electric potential* V, specified by the relations

$$H = \mu^{-1} \nabla \times A, \tag{14-1a}$$

$$D = i\omega[A + \beta \nabla \times A] - \nabla V. \tag{14-1b}$$

Furthermore, by using (14-1a,b) as well as the constitutive relations (3-5) it can be shown that

$$D = i\omega\varepsilon[A + 2\beta \nabla \times A + \beta^2 \nabla \times \nabla \times A] - \varepsilon\nabla V, \tag{14-1c}$$

$$B = \nabla \times [A + \beta\nabla \times A]. \tag{14-1d}$$

Next, by using (14-1a) and (14-1c) in the fourth Maxwell's equation, $\nabla \times H = -i\omega D$, the equation

$$\nabla^2 A + 2\gamma^2\beta \nabla \times A + \gamma^2 A - \nabla[\nabla \cdot A - i(\gamma^2/\omega)V] = 0 \tag{14-2}$$

can be obtained, where $\gamma^2 = k^2 [1-k^2\beta^2]^{-1}$ and $k^2 = \omega^2\mu\varepsilon$ as before. Provided the gauge condition,

$$i\omega\mu\varepsilon V - (k/\gamma)^2 \nabla \cdot A = 0, \tag{14-3}$$

is satisfied, A and V can be separated from each other and can be shown to satisfy the homogeneous governing differential equations

$$\nabla^2 A + 2\gamma^2\beta \nabla \times A + \gamma^2 A = 0, \tag{14-4a}$$

$$[\nabla^2 + \gamma^2] \, V = 0. \tag{14-4b}$$

In a similar fashion, the fourth Maxwell's equation, $\nabla \times H = -i\omega D$, can be completely satisfied by a *vector electric potential* F and a *scalar magnetic potential* W, which are specified by the relations

$$E = \varepsilon^{-1} \, \nabla \times F, \tag{14-5a}$$

$$H = -i\omega[F + \beta\nabla \times F] + \nabla W, \tag{14-5b}$$

$$B = -i\omega\mu[F + 2\beta \, \nabla \times F + \beta^2 \, \nabla \times \nabla \times F] + \mu\nabla W, \tag{14-5c}$$

$$D = \nabla \times [F + \beta\nabla \times F]. \tag{14-5d}$$

Again, provided the gauge condition,

$$i\omega\mu\varepsilon \, W - (k/\gamma)^2 \, \nabla \cdot F = 0 \tag{14-6}$$

holds, F and W can also be shown to satisfy the homogeneous governing differential equations

$$\nabla^2 F + 2\gamma^2\beta \, \nabla \times F + \gamma^2 F = 0, \tag{14-7a}$$

$$[\nabla^2 + \gamma^2] \, W = 0. \tag{14-7b}$$

Let now an electric current density J be impressed; then by using (14-1), $\nabla \times E = i\omega B$ and $\nabla \times H = -i\omega D + J$, it can be shown that the radiated magnetic potential A is governed by the relation

$$\nabla^2 A + 2\gamma^2\beta \, \nabla \times A + \gamma^2 A = -\mu(\gamma/k)^2 J. \tag{14-8}$$

Likewise, if a magnetic current density K be radiating then, Maxwell's equations read $\nabla \times E = i\omega B - K$ and $\nabla \times H = -i\omega D$; the radiated electric potential F has to be computed from the relation

$$\nabla^2 F + 2\gamma^2\beta \, \nabla \times F + \gamma^2 F = \varepsilon(\gamma/k)^2 K. \tag{14-9}$$

The solution of (14-8) as well as of (14-9) requires the derivation of an infinite-medium Green's dyadic $\mathfrak{A}(\mathbf{r},\mathbf{r}')$, which itself is the solution of

$$\left[\nabla^2\mathbf{I} + \gamma^2\mathbf{I} + 2\gamma^2\beta\,\nabla\times\mathbf{I}\right]\cdot\mathfrak{A}(\mathbf{r},\mathbf{r}') = -\mathbf{I}\delta(\mathbf{r}-\mathbf{r}').\tag{14-10}$$

In order to evaluate $\mathfrak{A}(\mathbf{r},\mathbf{r}')$, the three-dimensional (spatial) Fourier transforms

$$\mathfrak{A}(\mathbf{r},\mathbf{r}') = (2\pi)^{-3}\iiint_{-\infty}^{\infty} d^3p\,\mathfrak{a}(\mathbf{p})\,\exp[i\mathbf{p}\cdot(\mathbf{r}-\mathbf{r}')],\tag{14-11a}$$

$$\delta(\mathbf{r}-\mathbf{r}') = (2\pi)^{-3}\iiint_{-\infty}^{\infty} d^3p\,\exp[i\mathbf{p}\cdot(\mathbf{r}-\mathbf{r}')],\tag{14-11b}$$

are substituted in (14-10), yielding thereby the dyadic relation

$$\left[(\gamma^2-p^2)\,\mathbf{I} + 2i\gamma^2\beta\,\mathbf{p}\times\mathbf{I}\right]\cdot\mathfrak{a}(\mathbf{p}) = -\mathbf{I}.\tag{14-12}$$

Now, on noting that the inverse of the dyadic

$$\mathfrak{W} = \lambda\mathbf{I} + \mathbf{c}\times\mathbf{I}\tag{14-13a}$$

is given by

$$\mathfrak{W}^{-1} = [\lambda^2\mathbf{I} - \lambda\mathbf{c}\times\mathbf{I} + \mathbf{c}\mathbf{c}]/[\lambda^3 + \lambda\mathbf{c}\cdot\mathbf{c}],\tag{14-13b}$$

the solution of (14-12) can be found from dyadic algebra to be

$$\begin{aligned}\mathfrak{a}(\mathbf{p}) = \Big[&(p^2-\gamma^2)\,(p^2-\gamma_1{}^2)\,(p^2-\gamma_2{}^2)\,\Big]^{-1}\{-4\gamma^4\beta^2\}\,\mathbf{p}\mathbf{p}\\ &+ \Big[(p^2-\gamma_1{}^2)\,(p^2-\gamma_2{}^2)\,\Big]^{-1}\Big[(p^2-\gamma^2)\mathbf{I} + i2\gamma^2\beta\,\mathbf{p}\times\mathbf{I}\,\Big],\end{aligned}\tag{14-14}$$

where $\gamma_1 = k\,[1-k\beta]^{-1}$ and $\gamma_2 = \gamma^2/\gamma_1 = k\,[1+k\beta]^{-1}$ as before.

On taking the inverse Fourier transform of (14-14), *vide* (14-11a), it turns out that

$$8\pi^3\,\mathfrak{A}(\mathbf{r},\mathbf{r}') = \mathbf{I}\,K_1 + 2\gamma^2\beta\nabla\times\mathbf{I}\,K_2 + 4\gamma^4\beta^2\nabla\nabla\,K_3,\tag{14-15}$$

in which the integrals

$$K_1 = {}_{-\infty}\!\!\iiint^\infty d^3p \ (p^2-\gamma^2)\left[(p^2-\gamma_1{}^2)\,(p^2-\gamma_2{}^2)\,\right]^{-1} \exp[ip\cdot R]$$
$$= (2\pi/iR) \ {}_{-\infty}\!\!\int^\infty dp \ (p^2-\gamma^2)\left[(p^2-\gamma_1{}^2)\,(p^2-\gamma_2{}^2)\,\right]^{-1} p \ \exp[ipR], \quad (14\text{-}16a)$$

$$K_2 = {}_{-\infty}\!\!\iiint^\infty d^3p \ \left[(p^2-\gamma_1{}^2)\,(p^2-\gamma_2{}^2)\,\right]^{-1} \exp[ip\cdot R]$$
$$= (2\pi/iR) \ {}_{-\infty}\!\!\int^\infty dp \ \left[(p^2-\gamma_1{}^2)\,(p^2-\gamma_2{}^2)\,\right]^{-1} p \ \exp[ipR], \quad (14\text{-}16b)$$

$$K_3 = {}_{-\infty}\!\!\iiint^\infty d^3p \ \left[(p^2-\gamma^2)\,(p^2-\gamma_1{}^2)\,(p^2-\gamma_2{}^2)\,\right]^{-1} \exp[ip\cdot R]$$
$$= (2\pi/iR) \ {}_{-\infty}\!\!\int^\infty dp \ \left[(p^2-\gamma^2)\,(p^2-\gamma_1{}^2)\,(p^2-\gamma_2{}^2)\,\right]^{-1} p \ \exp[ipR], \quad (14\text{-}16c)$$

and $R = r - r'$. The evaluation of these integrals must be done in the upper half of the complex plane. To be noted is the fact that the integrands of both K_1 and K_2 contain singularities at $p = \pm\gamma_1, \pm\gamma_2$, while that of K_3 contains yet another singularity at $p = \pm\gamma$; of these six singularities, three have to be excluded because of the chosen time-dependence $\exp(-i\omega t)$. Therefore, after using Cauchy's residue theorem to evaluate the three integrals, the expression for the dyadic $\mathfrak{A}(r,r')$ turns out to be

$$\mathfrak{A}(r, r') = (k/8\pi\gamma^2) \left[\gamma_1\mathfrak{I} + \gamma_1{}^{-1}\,\nabla\nabla + \nabla\times\mathfrak{I}\right] g(\gamma_1;R),$$
$$+ (k/8\pi\gamma^2) \left[\gamma_2\mathfrak{I} + \gamma_2{}^{-1}\,\nabla\nabla - \nabla\times\mathfrak{I}\right] g(\gamma_2;R)$$
$$- (1/4\pi\gamma^2)\,\nabla\nabla g(\gamma;R), \quad (14\text{-}17)$$

where $g(\sigma;R) = \exp[i\sigma R]/R$. To be noted is the fact that when $\beta = 0$, then (14-17) simplifies to the usual expression for achiral media [Harrington 1964]

$$\mathfrak{A}(r, r') = (1/4\pi) \ \mathfrak{I} \ g(k;R). \quad (14\text{-}18)$$

Once the dyadic $\mathfrak{A}(r, r')$ has been derived, the radiated vector potentials can be easily evaluated from the integrals

$$A(r) = \mu(\gamma/k)^2 \int dv' \ \mathfrak{A}(r, r')\cdot J(r'), \quad (14\text{-}19a)$$
$$F(r) = -\varepsilon(\gamma/k)^2 \int dv' \ \mathfrak{A}(r, r')\cdot K(r'), \quad (14\text{-}19b)$$

in which the integrations are performed over the volumes containing the respective source current densities. The corresponding scalar potentials can then be obtained from the gauge conditions (14-3) and (14-6), if needed.

While the relation,

$$\mathfrak{A}(\mathbf{r},\mathbf{r}') = \mathfrak{G}(\mathbf{r}, \mathbf{r}') - (1/4\pi\gamma^2)\, \nabla\nabla g(\gamma;R), \tag{14-20a}$$

can be obtained by comparing (14-17) with the Green's function $\mathfrak{G}(\mathbf{r}, \mathbf{r}')$ of (13-13) for the electromagnetic field, it should be noted that

$$\nabla \times \mathfrak{A}(\mathbf{r},\mathbf{r}') = \nabla \times \mathfrak{G}(\mathbf{r}, \mathbf{r}'), \tag{14-20b}$$

because the second term on the right hand side of (14-20a) is irrotational.

Whereas the electromagnetic field vectors simply exhibit birefringence, it turns out to that the potential vectors, **A** and **F**, are trirefringent. As per (14-17), $\mathfrak{A}$ is comprised of three components, each having a different wavenumber -- γ_1, γ_2 and $\gamma = [\gamma_1\gamma_2]^{1/2}$, -- but all three being of the order $O(1/R)$ in the limit $R \to \infty$. The first two components of $\mathfrak{A}$ are solenoidal leading to the identity (14-20b); the third one, however, possessed with the phase velocity ω/γ, is purely longitudinal and does not contribute to the radiated **E** and **H**, which are solenoidal whatever be the value of β. Incidently, γ is a valid wavenumber for the potentials, as evinced by the homogeneous differential equations (14-4b) and (14-7b), to which, respectively, the scalar potentials V and W must conform. Finally, **A** and **F** are not axial vectors for $\beta \neq 0$; consequently, and because the solenoidal part of $\mathfrak{A}$ precisely equals $\mathfrak{G}$, the calculation of the radiated electromagnetic fields in isotropic chiral media through **A** and **F** is as simple or difficult as a direct calculation through $\mathfrak{G}$.

15. RADIATION IN CHIRAL MEDIA

The solution of the source-incorporated governing differential equations (4-10) for **E** and **H** fields in a Drude-Born-Fedorov medium is given by

$$(k/\gamma)^2 \mathbf{E}(\mathbf{r}) = i\omega\mu \int dv' \, [\mathbf{I} + \beta\nabla\times\mathbf{I}]\cdot\boldsymbol{\mathcal{B}}(\mathbf{r}, \mathbf{r}')\cdot\mathbf{J}(\mathbf{r}')$$
$$- \int dv' \, [\nabla\times\mathbf{I}]\cdot\boldsymbol{\mathcal{B}}(\mathbf{r}, \mathbf{r}')\cdot\mathbf{K}(\mathbf{r}'), \qquad (15\text{-}1a)$$
$$(k/\gamma)^2 \mathbf{H}(\mathbf{r}) = i\omega\epsilon \int dv' \, [\mathbf{I} + \beta\nabla\times\mathbf{I}]\cdot\boldsymbol{\mathcal{B}}(\mathbf{r}, \mathbf{r}')\cdot\mathbf{K}(\mathbf{r}')$$
$$+ \int dv' \, [\nabla\times\mathbf{I}]\cdot\boldsymbol{\mathcal{B}}(\mathbf{r}, \mathbf{r}')\cdot\mathbf{J}(\mathbf{r}'), \qquad (15\text{-}1b)$$

in which the integrations hold over the current-carrying volumes. After noting the rotational properties (13-16), it is possible to restate (15-1) in a form which clearly brings out the chiral flavor of the medium; *viz.*,

$$(k/\gamma)^2 \mathbf{E}(\mathbf{r}) = \gamma_1 \int dv' \, \boldsymbol{\mathcal{B}}_1(\mathbf{r}, \mathbf{r}')\cdot[(i\omega\mu/k)\mathbf{J}(\mathbf{r}') - \mathbf{K}(\mathbf{r}')] +$$
$$+ \gamma_2 \int dv' \, \boldsymbol{\mathcal{B}}_2(\mathbf{r}, \mathbf{r}')\cdot[(i\omega\mu/k)\mathbf{J}(\mathbf{r}') + \mathbf{K}(\mathbf{r}')], \qquad (15\text{-}2a)$$

$$(i\omega\mu/k) \, (k/\gamma)^2 \mathbf{H}(\mathbf{r}) = \gamma_1 \int dv' \, \boldsymbol{\mathcal{B}}_1(\mathbf{r}, \mathbf{r}')\cdot[(i\omega\mu/k)\mathbf{J}(\mathbf{r}') - \mathbf{K}(\mathbf{r}')] -$$
$$- \gamma_2 \int dv' \, \boldsymbol{\mathcal{B}}_2(\mathbf{r}, \mathbf{r}')\cdot[(i\omega\mu/k)\mathbf{J}(\mathbf{r}') + \mathbf{K}(\mathbf{r}')]. \qquad (15\text{-}2b)$$

Thus, if in a source volume $(i\omega\mu/k)\mathbf{J}(\mathbf{r}) + \mathbf{K}(\mathbf{r}) = \mathbf{0}$, then the radiated fields are purely LCP and $\mathbf{E}(\mathbf{r}) = (i\omega\mu/k)\mathbf{H}(\mathbf{r})$. On the other hand, should $(i\omega\mu/k)\mathbf{J}(\mathbf{r}) - \mathbf{K}(\mathbf{r}) = \mathbf{0}$, then the radiated fields are purely RCP and $\mathbf{E}(\mathbf{r}) = -(i\omega\mu/k)\mathbf{H}(\mathbf{r})$. Incidently, these are also precisely the canonical sources for the generation of the circularly polarized waves in isotropic *achiral* media, as pointed out by Rumsey [1961].

The equations (15-2) can be even more lucidly stated if the auxiliary fields $\mathbf{Q}_1$ and $\mathbf{Q}_2$ are used, where

$$\mathbf{Q}_1 = (1/2) \, [\mathbf{E} + (i\omega\mu/k)\mathbf{H}], \qquad (15\text{-}3a)$$
$$\mathbf{Q}_2 = (1/2)[(i\omega\epsilon/k)\mathbf{E} + \mathbf{H}]. \qquad (15\text{-}3b)$$

give the inverse of Bohren's transformation (7-1). In that case,

$$(k/\gamma)^2 \mathbf{Q}_1(\mathbf{r}) = \gamma_1 \int dv' \, \mathbf{G}_1(\mathbf{r}, \mathbf{r}') \cdot [(i\omega\mu/k)\mathbf{J}(\mathbf{r}') - \mathbf{K}(\mathbf{r}')], \qquad (15\text{-}4a)$$

$$(k/\gamma)^2 \mathbf{Q}_2(\mathbf{r}) = - (i\omega\mu/k) \, \gamma_2 \int dv' \, \mathbf{G}_2(\mathbf{r}, \mathbf{r}') \cdot [(i\omega\mu/k)\mathbf{J}(\mathbf{r}') + \mathbf{K}(\mathbf{r}')]. \quad (15\text{-}4b)$$

As can be observed from (15-4), the radiation of $\mathbf{Q}_1$ is independent from that of $\mathbf{Q}_2$. The coupling between the LCP and the RCP fields takes place only at bimaterial boundaries [Lakhtakia et al 1985, 1987a] where conditions on the tangential components of $\mathbf{E}$ and $\mathbf{H}$ must be satisfied, i.e., the boundary conditions are specified not on the $\mathbf{Q}$'s singly, but on the tangential components of the combinations $[\mathbf{Q}_1 - (i\omega\mu/k)\mathbf{Q}_2]$ and $[\mathbf{Q}_2 - (i\omega\varepsilon/k)\mathbf{Q}_1]$.

The fields radiated by current sources in chiral media are different from those in achiral ones. To see that, let an electric dipole source $\mathbf{p}$ be located at the origin. In this case, $\mathbf{J}(\mathbf{r}) = -i\omega\mathbf{p}\delta(\mathbf{r})$, and from (15-1) the radiated fields can be derived to be

$$\mathbf{E}(\mathbf{r}) = (\omega^2\mu/k) \, (\gamma/k)^2 \, [\gamma_1 \mathbf{G}_1(\mathbf{r}, 0) + \gamma_2 \mathbf{G}_2(\mathbf{r}, 0)] \cdot \mathbf{p}, \qquad (15\text{-}5a)$$

$$\mathbf{H}(\mathbf{r}) = -i\omega \, (\gamma/k)^2 \, [\gamma_1 \mathbf{G}_1(\mathbf{r}, 0) - \gamma_2 \mathbf{G}_2(\mathbf{r}, 0)] \cdot \mathbf{p}. \qquad (15\text{-}5b)$$

Likewise, for a magnetic dipole $\mathbf{m}$ located at the origin, $\mathbf{J}(\mathbf{r}) = \nabla \times [\mathbf{m}\delta(\mathbf{r})]$, and the radiated fields turn out to be

$$\mathbf{E}(\mathbf{r}) = (i\omega\mu/k) \, (\gamma/k)^2 \, [\gamma_1^2 \mathbf{G}_1(\mathbf{r}, 0) - \gamma_2^2 \mathbf{G}_2(\mathbf{r}, 0)] \cdot \mathbf{m}, \qquad (15\text{-}6a)$$

$$\mathbf{H}(\mathbf{r}) = (\gamma/k)^2 \, [\gamma_1^2 \mathbf{G}_1(\mathbf{r}, 0) + \gamma_2^2 \mathbf{G}_2(\mathbf{r}, 0)] \cdot \mathbf{m}. \qquad (15\text{-}6b)$$

The important difference between chiral and achiral media can be easily seen now by examining (15-5) and (15-6). Without loss of generality, let the source dipole moments be parallel to the z axis. Then, if the dipole moments are radiating in an achiral medium, at $\mathbf{r} = z\mathbf{e}_z$ there is no H-field due to $\mathbf{p}$ and there is no E-field due to $\mathbf{m}$. On the other hand, the phase differences between the LCP and RCP components guarantee that, in a chiral medium, both $\mathbf{E}$ and $\mathbf{H}$ fields exist on the z axis regardless of which dipole moment is radiating.

A similar conclusion can be drawn if the source is a constant-current loop of radius a. Therefore,

$$\mathbf{J}(\mathbf{r}) = (I_o/a)\mathbf{e}_\varphi \, \delta(r\text{-}a) \, \delta(\theta\text{-}\pi/2). \qquad (15\text{-}7)$$

Using (15-1), the radiation field on the z axis can be worked out to yield

$$\mathbf{E}(z\mathbf{e}_z) = \mathbf{e}_z \, (kI_o\pi a^2)(i\omega\mu/k)[(\gamma_1/k)^2 h(\gamma_1;R_a) - (\gamma_2/k)^2 h(\gamma_2;R_a)], \qquad (15\text{-}8a)$$

$$\mathbf{H}(z\mathbf{e}_z) = \mathbf{e}_z \, (kI_o\pi a^2) \, [(\gamma_1/k)^2 h(\gamma_1;R_a) + (\gamma_2/k)^2 h(\gamma_2;R_a)], \qquad (15\text{-}8b)$$

in which $h(\sigma;R) = [(iR)^{-1} + (\sigma R^2)^{-1}]g(\sigma;R)$ and $R_a = \sqrt{[a^2 + z^2]}$. It is easy to see that if $\beta = 0$, then the right side of (15-8a) reduces identically to zero, as would be expected.

Let now a more complex source be considered: the straight-wire antenna shown in Fig. 15-1, whose current-density profile is given by

$$\mathbf{J}(\mathbf{r}) = \mathbf{e}_z \, I_o \, f(z) \, \delta(x) \, \delta(y) \, [U(z+L) - U(z-L)], \qquad (15\text{-}9)$$

where $U(\cdot)$ is the Heaviside function, $f(z)$ is some bounded function which satisfactorily describes the conditions prevalent at the feed point and the open ends of the antenna, and I_o is a constant. Although very cumbersome, expressions for the radiated $\mathbf{E}$ and $\mathbf{H}$ fields can now be derived by substituting the relations

$$\nabla\nabla g(\sigma;R) = \left[\{i\sigma R^{-1} - R^{-2}\}(\mathbf{\mathcal{I}} - 3\,\mathbf{RR}\,R^{-2}) - (\sigma/R)^2\,\mathbf{RR} \right] g(\sigma;R), \quad (15\text{-}10a)$$

$$\nabla\times[\mathbf{\mathcal{I}}g(\sigma;R)] = \{i\sigma R^{-1} - R^{-2}\} \, g(\sigma;R) \, \mathbf{R}\times\mathbf{\mathcal{I}}, \qquad (15\text{-}10b)$$

along with (15-9) into (15-1); also

$$\nabla\times\mathbf{J}(\mathbf{r}) = I_o \, f(z)[U(z+L) - U(z-L)]\{\mathbf{e}_x \, \delta(x) \, \delta'(y) - \mathbf{e}_y \, \delta'(x) \, \delta(y)\}, \quad (15\text{-}11)$$

in which $\delta'(y) = d\delta'(y)/dy$, etc.

More pertinent for the solution of this radiation problem are the values of the radiation fields in the far zone. On retaining only the lowest order terms in r^{-1}, we arrive at the asymptotic expansion of the Green's dyadic in the far zone as

$$\mathbf{\Phi}_1(\mathbf{r}, \mathbf{r}') = (ik\gamma_1/8\pi\gamma^2) \, \mathbf{e}_r \times \left[i\mathbf{e}_r \times \mathbf{\mathcal{I}} + \mathbf{\mathcal{I}} \right] g(\gamma_1;r) \, \exp[-i\gamma_1\mathbf{e}_r\cdot\mathbf{r}'], \qquad (15\text{-}12a)$$

$$\mathbf{\Phi}_2(\mathbf{r}, \mathbf{r}') = (ik\gamma_2/8\pi\gamma^2) \, \mathbf{e}_r \times \left[i\mathbf{e}_r \times \mathbf{\mathcal{I}} - \mathbf{\mathcal{I}} \right] g(\gamma_2;r) \, \exp[-i\gamma_2\mathbf{e}_r\cdot\mathbf{r}']. \qquad (15\text{-}12b)$$

Next, after using (15-9), (15-11), (15-12) as well as the properties of the Dirac delta function in (15-2), the far-zone radiated fields radiated by the antenna of Fig. 15-1 can be compactly set down as

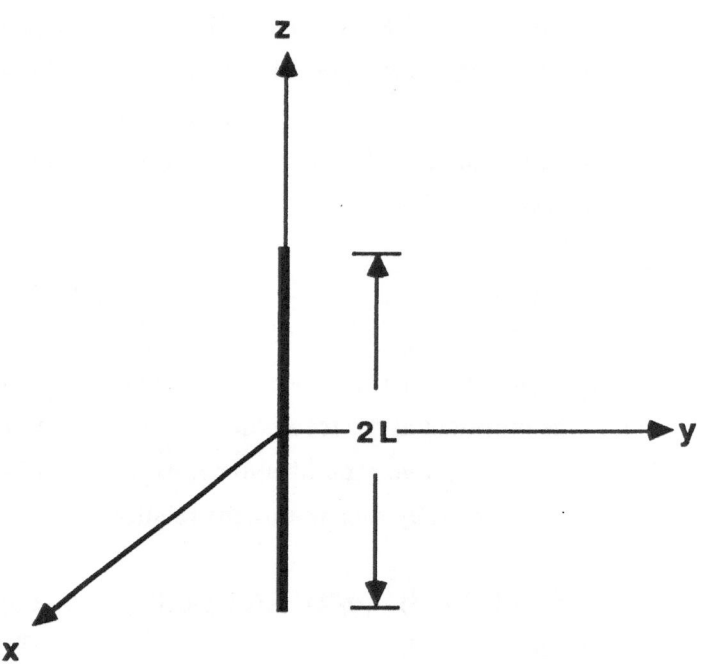

Fig. 15-1 Relevant to radiation by a straight thin-wire antenna in a Drude-Born-Fedorov medium

$$\mathbf{E}(\mathbf{r}) = (-i\omega\mu I_o/8\pi)\ \sin\theta\left\{(\mathbf{e_\theta}+i\mathbf{e_\phi})(\gamma_1/k)^2\ \mathbb{P}[\gamma_1\cos\theta]g(\gamma_1;r)\right.$$
$$\left. + (\mathbf{e_\theta}-i\mathbf{e_\phi})(\gamma_2/k)^2\ \mathbb{P}[\gamma_1\cos\theta]\ g(\gamma_2;r)\right\}, \tag{15-13a}$$

$$\mathbf{H}(\mathbf{r}) = (-kI_o/8\pi)\ \sin\theta\left\{(\mathbf{e_\theta}+i\mathbf{e_\phi})(\gamma_1/k)^2\ \mathbb{P}[\gamma_1\cos\theta]g(\gamma_1;r)\right.$$
$$\left. - (\mathbf{e_\theta}-i\mathbf{e_\phi})(\gamma_2/k)^2\ \mathbb{P}[\gamma_1\cos\theta]\ g(\gamma_2;r)\right\}, \tag{15-13b}$$

with the integral $\mathbb{P}[\sigma]$ defined by

$$\mathbb{P}[\sigma]\ =\ {}_{-L}\!\!\int^L dz_o\ f(z_o)\ \exp[-i\sigma z_o]. \tag{15-14}$$

Needless to add, should $\beta = 0$, then (15-13a,b) simplify, respectively, to

$$\mathbf{E}(\mathbf{r}) = (-i\omega\mu I_o/4\pi) \sin\theta \, g(k;r) \, \mathbb{P}[k\cos\theta] \, \mathbf{e}_\theta, \qquad (15\text{-}15a)$$

$$\mathbf{H}(\mathbf{r}) = (-ikI_o/4\pi) \sin\theta \, g(k;r) \, \mathbb{P}[k\cos\theta] \, \mathbf{e}_\varphi, \qquad (15\text{-}15b)$$

as expected. Parenthetically, it should be noted that (15-14) defines a spatial Fourier transform of the kind expected from Fraunhoffer diffraction in the optics of achiral media.

It is easy to see from (15-13) that the far-zone radiated field is elliptically-polarized, with the terms containing $(\mathbf{e}_\theta+i\mathbf{e}_\varphi)$ representing the left-circularly polarized (LCP) waves, and the terms containing $(\mathbf{e}_\theta-i\mathbf{e}_\varphi)$ representing the right-circularly polarized (RCP) waves. Furthermore, neither of the fields contains a radial component, indicating that the far-zone radiation is transverse electromagnetic (TEM) in character; and the radiation pattern is symmetric about the plane $\theta = \pi/2$. To be noted also is the fact that $\mathbf{E}$ and $\mathbf{H}$ are zero along the z-axis and the radiation pattern is azimuthally isotropic, both of these features also to be found for isotropic achiral media.

In an achiral medium, there is only one wavenumber; hence the spherical wavefronts of the radiation field are only trivially dependent on r *via* the factor exp[ikr]/r. The wavefronts are still spherical for the chiral medium; however, because of the birefringence of this medium, attention is directed to the constant spherical surfaces $r = n\pi/2\beta\gamma^2$ in the far zone, where n is a positive integer. It is easy to see from (15-13a) that

$$E_{LCP}/E_{RCP} = \{\mathbb{P}[\gamma_1\cos\theta]/\,\mathbb{P}[\gamma_2\cos\theta]\}[1+k\beta]^2 \, [1-k\beta]^{-2}; \, n \; even, \qquad (15\text{-}16a)$$

$$E_{LCP}/E_{RCP} = -\{\mathbb{P}[\gamma_1\cos\theta]/\,\mathbb{P}[\gamma_2\cos\theta]\}[1+k\beta]^2 \, [1-k\beta]^{-2}; \, n \; odd, \qquad (15\text{-}16b)$$

showing us that the LCP and the RCP components of the radiated field keep on flipping signs as one moves over the successive spherical surfaces. Furthermore, and perhaps more importantly,

$$E_\theta/E_\varphi = -i\{(\gamma_1/k)^2 \, \mathbb{P}[\gamma_1\cos\theta] - (\gamma_2/k)^2 \, \mathbb{P}[\gamma_2\cos\theta]\}^{-1}$$
$$\cdot \{(\gamma_1/k)^2 \, \mathbb{P}[\gamma_1\cos\theta] + (\gamma_2/k)^2 \, \mathbb{P}[\gamma_2\cos\theta]\}; \, n \; even, \qquad (15\text{-}18a)$$

$$E_\theta/E_\varphi = -i\{(\gamma_1/k)^2 \, \mathbb{P}[\gamma_1\cos\theta] - (\gamma_2/k)^2 \, \mathbb{P}[\gamma_2\cos\theta]\}$$
$$\bullet \{(\gamma_1/k)^2 \, \mathbb{P}[\gamma_1\cos\theta] + (\gamma_2/k)^2 \, \mathbb{P}[\gamma_2\cos\theta]\}^{-1}; \text{ n } \textit{odd}, \quad (15\text{-}18\text{b})$$

showing that E_θ *leads* E_φ if the two factors within the braces in (15-18) are both of the same sign, and E_θ *lags* E_φ if the two terms have differing signs -- this regardless of the specific value of n. If the antenna excitation function $f(z) = \delta(z)$, then $\mathbb{P}[\bullet] = 1$, and (15-18a,b) simplify considerably to

$$E_\theta/E_\varphi = -i \, [2k\beta]^{-1} \, [1+k^2\beta^2]; \text{ n } \textit{even}, \qquad\qquad (15\text{-}19\text{a})$$
$$E_\theta/E_\varphi = -i \, [2k\beta] \, [1+k^2\beta^2]^{-1}; \text{ n } \textit{odd}. \qquad\qquad (15\text{-}19\text{b})$$

This shows that for $\beta > 0$ (resp. $\beta < 0$), E_θ (E_φ) *leads* E_φ (E_θ) irrespective of the specific spherical surface we focus our attention on. Moreover, with the usual condition that $|k\beta| < 1$, it turns out that $|E_\theta| > |E_\varphi|$ when n is *even*, and $|E_\theta| < |E_\varphi|$ if otherwise.

The total power P_{rad}, radiated by the antenna of Fig. 15-1 can be computed via (15-13) and turns out to be

$$P_{rad} = Re \left(\{k\omega\mu/32\pi\} \, |I_0|^2 \, {}_0\!\!\int^\pi d\theta \, \sin^3\theta \right.$$
$$\left. \bullet \{(\gamma_1/k)^4 \, |\mathbb{P}[\gamma_1\cos\theta]|^2 + (\gamma_2/k)^4 \, |\mathbb{P}[\gamma_2\cos\theta]|^2\} \right). \quad (15\text{-}20)$$

Evaluation of the integral involves the prescription of $f(z)$; nevertheless, an important characteristic of the isotropic chiral media can be deduced by letting the antenna be a point electric dipole, i.e., $f(z) = \delta(z)$. In that case,

$$P_{rad} = \{k\omega\mu/24\pi\} \, |I_0|^2 \, \{(1-k\beta)^{-4} + (1+k\beta)^{-4}\}. \qquad (15\text{-}21\text{a})$$

If the radiation resistance, R, is defined by

$$R = 2 \, |I_0|^{-2} \, P_{rad}, \qquad\qquad (15\text{-}21\text{b})$$

then, for point dipole antennas, it is easy to see that the ratio

$$R/R_{\beta=0} = [1+ 6(k\beta)^2 + (k\beta)^4] [1 - (k\beta)^2]^{-4}, \tag{15-21c}$$

grows monotonically with $(k\beta)^2$ for $|k\beta| < 1$. This means that the incorporation of chirality in an otherwise achiral medium enhances the radiation resistance. Parenthetically, it is pointed out that if $k\beta = \pm 1$, this ratio blows up; one generally expects $|k\beta| \ll 1$.

Furthermore, if $f(z) = \delta(z)$, the total gain of the point dipole antenna still remains 1.5 which would also be the case for achiral media [Johnk 1975] as well. The ratio of the LCP part of the gain to the RCP part works out to be

$$G_{LCP}/G_{RCP} = (\gamma_1/\gamma_2)^4. \tag{15-22}$$

In a right-handed medium, $\gamma_1 > \gamma_2$, implying that G_{LCP} is the higher of the two; the reverse being the case for left-handed media.

What is the maximum achievable gain from any antenna radiating into a chiral medium? Outside a minimum sphere circumscribing the antenna, the radiated fields can be expressed in the series expansion

$$\mathbf{E} = \Sigma_{v=smn} \, c_v \, \mathbf{L}_v^{(3)}(\mathbf{r}) + a_R \, \Sigma_{v=smn} \, d_v \, \mathbf{R}_v^{(3)}(\mathbf{r}), \tag{15-23a}$$
$$\mathbf{H} = a_L \, \Sigma_{v=smn} \, c_v \, \mathbf{L}_v^{(3)}(\mathbf{r}) + \Sigma_{v=smn} \, d_v \, \mathbf{R}_v^{(3)}(\mathbf{r}), \tag{15-23b}$$

as per (15-2) and Sections 12 and 13, in which $v = smn$ is the triple-index mentioned for (12-6). The coefficients c_v and d_v have to be adjusted in order to obtain the maximum gain.

The total gain G of the antenna is defined by

$$G = 4\pi \, r^2 \, (S_r)_{max}/P_{rad}, \tag{15-24}$$

in which P_{rad} is the total time-averaged power and $(S_r)_{max}$ is the maximum radial power density. The total time-averaged power radiated by the antenna can be calculated as

$$P_{rad} = (1/2) \, \Sigma_{\sigma mn} \, \{[8\pi/(1+\delta_{m0})(2n+1)][(n+1)! \, (n+m)!]/[(n-1)! \, (n-m)!] \cdot$$

$$[(1/\varpi \ \gamma_1{}^2) \ |c_v|^2 + (\varpi/\gamma_2{}^2) \ |d_v|^2 \ ], \qquad (15\text{-}25)$$

where $\varpi = \sqrt{(\mu/\epsilon)}$. It is understood here that the medium is lossless and that the $\{s = \text{odd}, m = 0\}$ modes do not exist.

Let the radial power density S_r be maximum in the direction $\theta = 0$. Then, because only the $\{m = 1\}$ modes contribute to this direction,

$$r^2 \ (S_r)_{max} = (1/\varpi \ \gamma_1{}^2) \ |\Sigma_{sn} \ c_{s1n} \ (\text{-}i)^{n+1} \ n(n+1)/2|^2$$
$$+ (\varpi/\gamma_2{}^2) \ |\Sigma_{sn} \ d_{s1n} \ (\text{-}i)^{n+1} \ n(n+1)/2|^2, \qquad (15\text{-}26)$$

in the far zone.

It can be seen from (15-24) that G can be maximized by minimizing P_{rad}. Thus, one condition for maximizing G can be easily found as

$$c_{smn} = d_{smn} = 0 \ \ \forall m \neq 1. \qquad (15\text{-}27)$$

Consequently, with the help of the normalized coefficients

$$\rho_{sn} = c_{s1n} \ (\text{-}i)^{n+1} \ n(n+1)/2\gamma_1, \qquad (15\text{-}28a)$$
$$\tau_{sn} = \varpi \ d_{s1n} \ (\text{-}i)^{n+1} \ n(n+1)/2\gamma_2, \qquad (15\text{-}28b)$$

the gain turns out to be

$$G = (1/2) \ \left[|\Sigma_{sn} \ \rho_{sn}|^2 + |\Sigma_{sn} \ \tau_{sn} \ |^2\right] / \left[\Sigma_{sn} \ (2n+1)^{-1}\{|\rho_{sn}|^2 + |\tau_{sn}|^2\}\right]. \quad (15\text{-}29)$$

The denominator of the right side of (15-29) is independent of the phases of ρ_{sn} and τ_{sn}; therefore, the gain can be further maximized through the maximization of the numerator on the right side of (15-29) by having

$$Im \ \rho_{sn} = Im \ \tau_{sn} = 0 \ \ \ \forall\{s,n\}. \qquad (15\text{-}30)$$

The gain given by (15-29), subject to the conditions (15-27) and (15-30), is unbounded so long as n itself is unbounded. If however the field specified by

(2a,b) contains only wavefunctions of order $1 \leq n \leq N$, then an upper limit of G exists. In that case, G will be at its maximum provided

$$\partial G / \partial \rho_{s'n'} = \partial G / \partial \tau_{s'n'} = 0 \qquad \forall \{s', 1 \leq n' \leq N\}. \tag{15-31}$$

On noting that ρ_{sn} and τ_{sn} are independent of the parity index s, it can be easily shown that [Harrington 1964]

$$\rho_{sn} = (2n+1) \rho_{e1}/3; \qquad \tau_{sn} = (2n+1) \tau_{e1}/3 \qquad \forall \{s, 1 \leq n \leq N\}, \tag{15-32a}$$

with both ρ_{e1} and τ_{e1} real are needed to obtain the maximum possible gain. Under these conditions, the total radiated power is given by

$$P_{rad} = (16\pi/9\varpi) \ N(N+2) \ \{\rho_{e1}{}^2 + \tau_{e1}{}^2\}, \tag{15-33a}$$

while the maximum radial power density can be obtained from

$$r^2 \ (S_r)_{max} = (4/9\varpi) \ N^2 \ (N+2)^2 \ \{\rho_{e1}{}^2 + \tau_{e1}{}^2\}. \tag{15-33b}$$

The value of $G_{max} = N(N+2)$, obtained by substituting (15-33) into (15-24), has also been derived by Harrington [1964] for an antenna radiating into an achiral medium. Nevertheless, it should be noted that the radiation pattern of any antenna radiating into an achiral medium is different from that of the same antenna radiating into a chiral medium.

16. EQUIVALENCE OF SOURCES

It is often possible that a field problem is simplified by replacing electric sources with equivalent magnetic sources, and *vice versa* [Harrington 1964]. To that end, for homogeneous, achiral, isotropic media the necessary source-equivalence theorems were derived by Mayes [1958]; the same can also be accomplished for isotropic chiral media.

Let $\{E_1, H_1\}$ be the fields produced by an electric current density J, while a magnetic current distribution K independently creates the fields $\{E_2, H_2\}$. From (4-10a,b) it follows that

$$\mathbf{D} \cdot (E_1 - E_2) = (\gamma/k)^2 \left(i\omega\mu[J + \beta\nabla\times J] + \nabla\times K \right), \qquad (16\text{-}1a)$$

$$\mathbf{D} \cdot (H_1 - H_2) = (\gamma/k)^2 \left(\nabla\times J - i\omega\varepsilon[K + \beta\nabla\times K] \right), \qquad (16\text{-}1b)$$

where the dyadic operator

$$\mathbf{D} = \nabla\times\nabla\times\mathbf{I} - 2\gamma^2\beta \, \nabla\times\mathbf{I} - \gamma^2 \, \mathbf{I} \qquad (16\text{-}2)$$

has been used for convenience. If the two sources are such that

$$J + \beta\nabla\times J = -\nabla\times K/i\omega\mu, \qquad (16\text{-}3)$$

then (16-1a) yields the identity $E_1 = E_2$. At the same time, from (4-2b) and by enforcing the provision that $E_1 - E_2 = 0$, it can be seen that

$$\nabla\times(H_1 - H_2) = \gamma^2\beta \, (H_1 - H_2) + (\gamma/k)^2 \, [J - i\omega\varepsilon\beta K], \qquad (16\text{-}4a)$$

wherefore,

$$\nabla\times\nabla\times (H_1 - H_2) = \gamma^2\beta\nabla\times (H_1 - H_2) + (\gamma/k)^2 \, [\nabla\times J - i\omega\varepsilon\beta\nabla\times K]. \quad (16\text{-}4b)$$

Substitution of (16-4b) in (16-1b), and the subsequent use of the constitutive equation (3-5b), then yields the identity $B_1 - B_2 = -K/i\omega$. It should be noted here that the specification of an E-equivalent $\mathbf{K}$ for a specified $\mathbf{J}$ is non-unique to the extent that $\mathbf{K} \rightarrow \mathbf{K} + \nabla\zeta$, $\zeta(\mathbf{r})$ being any arbitrary scalar field. This non-uniqueness of $\mathbf{K}$ does not affect the computation of the radiated electric field; as can be seen from (15-1a),

$$(k/\gamma)^2 \, \mathbf{E}_2(\mathbf{r}) = - \int dv' \, [\nabla\times\mathbf{1}]\cdot\mathbf{\mathcal{G}}(\mathbf{r}, \, \mathbf{r}')\cdot\mathbf{K}(\mathbf{r}'), \qquad (16\text{-}5)$$

which can in no way be influenced by the replacement of $\mathbf{K}$ by $\mathbf{K} + \nabla\zeta$; $\mathbf{r}$ and $\mathbf{r}'$, respectively, are the field and the source points with $\mathbf{r}$ lying outside the source-carrying volume. It follows then also that $\mathbf{D}_1(\mathbf{r}) = \mathbf{D}_2(\mathbf{r})$. On the the other hand, the differences $\mathbf{B}_1(\mathbf{r}) - \mathbf{B}_2(\mathbf{r})$ and $\mathbf{H}_1(\mathbf{r}) - \mathbf{H}_2(\mathbf{r})$ are neither unique nor zero, but that consideration is not required here.

Proceeding in the same way, and with similar considerations, it can also be shown if two independent source current distributions $\mathbf{J}$ and $\mathbf{K}$ are such that

$$\mathbf{K} + \beta\nabla\times\mathbf{K} = \nabla\times\mathbf{J}/i\omega\varepsilon, \qquad (16\text{-}6)$$

then they produce identical magnetic fields, i.e., $\mathbf{H}_1 = \mathbf{H}_2$, while $\mathbf{D}_1 - \mathbf{D}_2 = \mathbf{J}/i\omega$. This prescription of an H-equivalent $\mathbf{J}$, is non-unique to the extent that $\mathbf{J} \rightarrow \mathbf{J} + \nabla\xi$, $\xi(\mathbf{r})$ being any arbitrary scalar field.

Of great interest would be finding $\mathbf{J}$ and $\mathbf{K}$ such that both conditions (16-3) and (16-6) are simultaneously satisfied, i.e., $\mathbf{E}_1 = \mathbf{E}_2$ along with $\mathbf{H}_1 = \mathbf{H}_2$. By substituting for $\beta\nabla\times\mathbf{K}$ from (16-6) into (16-3), it is easy to see that $i\omega\varepsilon\mathbf{K} = (k/\gamma)^2\nabla\times\mathbf{J} - k^2\beta\mathbf{J}$. Next, the curl of this equation is taken, and $\nabla\times\mathbf{K}$ from (16-3) is substituted. As a result, $\mathbf{\mathcal{D}}\cdot\mathbf{J} = \mathbf{0}$. Furthermore, by substituting for $\nabla\times\mathbf{K}$ from (16-3) into (16-6), it can be seen that $\mathbf{K} = k^2(i\omega\varepsilon)^{-1}[\nabla\times\mathbf{J}/\gamma^2 - \beta\mathbf{J}]$. Thus, if

$$\mathbf{\mathcal{D}}\cdot\mathbf{J} = \mathbf{0}, \qquad (16\text{-}7a)$$

$$\mathbf{K} = k^2(i\omega\varepsilon)^{-1}[\nabla\times\mathbf{J}/\gamma^2 - \beta\mathbf{J}], \qquad (16\text{-}7b)$$

then they produce identical E and H fields.

In order to show that the $\mathbf{J}$ and $\mathbf{K}$ of (16-7) produce the same E and H fields, consider (4-10a) and (16-7b), whence

$$\mathbf{B} \cdot \mathbf{E}_2 = -(\gamma/k)^2 \nabla \times \mathbf{K} = -(i\omega\varepsilon)^{-1}[\nabla \times \nabla \times \mathbf{J} - \gamma^2 \beta \nabla \times \mathbf{J}]; \qquad (16\text{-}8\text{a})$$

on using now (16-7a), this can be simplified to

$$\mathbf{B} \cdot \mathbf{E}_2 = (i\omega\mu)(\gamma/k)^2 \, [\mathbf{J} + \beta \nabla \times \mathbf{J}] = \mathbf{B} \cdot \mathbf{E}_1. \qquad (16\text{-}8\text{b})$$

Likewise, from (4-10b) and (16-7), it can be shown that

$$\mathbf{B} \cdot \mathbf{H}_2 = (i\omega\varepsilon)(\gamma/k)^2 \, [\mathbf{K} + \beta \nabla \times \mathbf{K}] = (\gamma/k)^2 \, [\nabla \times \mathbf{J}] = \mathbf{B} \cdot \mathbf{H}_1. \qquad (16\text{-}9)$$

Finally comes the question of the integrity of the fields $\{\mathbf{E}_1, \mathbf{H}_1\}$ radiated by $\mathbf{J}$. Then, after using the transpose properties (13-15) of the Green's dyadic, from (15-1) one has at a field point $\mathbf{r}$ where $\mathbf{J}(\mathbf{r}) = 0$,

$$\mathbf{E}_1(\mathbf{r}) = (i\omega\mu)(\gamma/k)^2 \int dv' \mathbf{B}(\mathbf{r}, \mathbf{r}') \cdot [\mathbf{J}(\mathbf{r}') + \beta \nabla' \times \mathbf{J}(\mathbf{r}')\,], \qquad (16\text{-}10\text{a})$$

$$\nabla \times \mathbf{E}_1(\mathbf{r}) = (i\omega\mu)(\gamma/k)^2 \int dv' \; \mathbf{B}(\mathbf{r}, \mathbf{r}') \cdot [\nabla' \times \mathbf{J}(\mathbf{r}') + \beta \nabla' \times \nabla' \times \mathbf{J}(\mathbf{r}')], \quad (16\text{-}10\text{b})$$

$$\mathbf{H}_1(\mathbf{r}) = (\gamma/k)^2 \int dv' \; \mathbf{B}(\mathbf{r}, \mathbf{r}') \cdot [\nabla' \times \mathbf{J}(\mathbf{r}')], \qquad (16\text{-}10\text{c})$$

$$\nabla \times \mathbf{H}_1(\mathbf{r}) = (\gamma/k)^2 \int dv' \; \mathbf{B}(\mathbf{r}, \mathbf{r}') \cdot [\nabla' \times \nabla' \times \mathbf{J}(\mathbf{r}')]. \qquad (16\text{-}10\text{d})$$

Together these relations imply that

$$\begin{aligned}
\nabla \times \mathbf{E}_1(\mathbf{r}) &- i\omega \mathbf{B}_1(\mathbf{r}) \\
&= \nabla \times \mathbf{E}_1(\mathbf{r}) - i\omega\mu \mathbf{H}_1(\mathbf{r}) - i\omega\mu\beta \nabla \times \mathbf{H}_1(\mathbf{r}) \\
&= (i\omega\mu) \, (\gamma/k)^2 \int dv' \; \mathbf{B}(\mathbf{r}, \mathbf{r}') \cdot [\nabla' \times \mathbf{J}(\mathbf{r}') + \beta \nabla' \times \nabla' \times \mathbf{J}(\mathbf{r}') \\
&\qquad\qquad\qquad\qquad\qquad - \nabla' \times \mathbf{J}(\mathbf{r}') - \beta \nabla' \times \nabla' \times \mathbf{J}(\mathbf{r}')] \\
&= 0, \qquad\qquad\qquad\qquad\qquad\qquad\qquad\qquad\qquad\qquad\qquad (16\text{-}11\text{a})
\end{aligned}$$

as is appropriate at the source-free point $\mathbf{r}$, *vide* Faraday's law. In a similar fashion, from (16-7a) and (16-10), it can also be shown that

$$\nabla \times \mathbf{H}_1(\mathbf{r}) + i\omega \mathbf{D}_1(\mathbf{r}) = 0. \qquad (16\text{-}11\text{b})$$

Thus, from (16-11) it is clear that the fields radiated by $\mathbf{J}$ of (16-7a) are fully compatible with Maxwell's equations.

It should also be noted that if there exists a magnetic current density $\mathbf{K}$ such that

$$\nabla \cdot \mathbf{K} = 0, \tag{16-12a}$$

then there exists also an electric current density $\mathbf{J}$, given by

$$\mathbf{J} = -k^2(i\omega\mu)^{-1}[\nabla \times \mathbf{K}/\gamma^2 - \beta\mathbf{K}], \tag{16-12b}$$

such that they both produce the same electric and magnetic fields, and also satisfy Maxwell's equations.

17. HUYGENS'S PRINCIPLE AND SCATTERING FORMALISMS

In order to investigate the scattering response of an obstacle embedded in a chiral medium, it is necessary to formulate Huygens's principle for such media. Therefore, consider the sourceless volume V_e bounded by the surfaces S and S_∞. Using Green's second identity for vectors $\mathbf{U}$ and $\mathbf{W} = \mathcal{G}(\mathbf{r}, \mathbf{r}') \cdot \mathbf{a}$ in V_e, one obtains

$$\int_{Ve} dv \left\{ \mathbf{U} \cdot (\nabla \times \nabla \times \mathbf{W}) - (\nabla \times \nabla \times \mathbf{U}) \cdot \mathbf{W} \right\}$$
$$= \int_{S+S\infty} ds \left\{ \mathbf{e}_n \times (\nabla \times \mathbf{U}) \cdot \mathbf{W} + (\mathbf{e}_n \times \mathbf{U}) \cdot (\nabla \times \mathbf{W}) \right\}, \qquad (17\text{-}1)$$

where $\mathbf{e}_n$ is the unit normal shown in Fig.17-1. By letting $\mathbf{a}$ be a constant vector, this equation can be stated in a more relevant form as

$$\int_{Ve} dv \left\{ \mathbf{U} \cdot (\nabla \times \nabla \times \mathcal{G}) - (\nabla \times \nabla \times \mathbf{U}) \cdot \mathcal{G} \right\}$$
$$= \int_{S+S\infty} ds \left\{ \mathbf{e}_n \times (\nabla \times \mathbf{U}) \cdot \mathcal{G} + (\mathbf{e}_n \times \mathbf{U}) \cdot (\nabla \times \mathcal{G}) \right\}. \qquad (17\text{-}2)$$

It is assumed that in V_e there are no sources; consequently, *vide* (4-7),

$$(\nabla \times \nabla \times \mathbf{U}) = \gamma^2 [\mathbf{U} + 2\beta \nabla \times \mathbf{U}], \qquad (17\text{-}3a)$$

and

$$(\nabla \times \nabla \times \mathcal{G}) = \gamma^2 [\mathcal{G} + 2\beta \nabla \times \mathcal{G}] + \mathbf{I} \delta(\mathbf{r} - \mathbf{r}'), \qquad (17\text{-}3b)$$

from (13-5). On substituting these two relations in (17-2), it is seen that

$$\int_{Ve} dv \left\{ \mathbf{U}(\mathbf{r}) \cdot \mathbf{I} \delta(\mathbf{r} - \mathbf{r}') \right\} = -2\gamma^2 \beta \int_{S+S\infty} ds\, \mathbf{e}_n \cdot [\mathbf{U}(\mathbf{r}) \times \mathcal{G}(\mathbf{r}, \mathbf{r}')]$$
$$+ \int_{S+S\infty} ds \left\{ \mathbf{e}_n \times [\nabla \times \mathbf{U}(\mathbf{r})] \cdot \mathcal{G}(\mathbf{r}, \mathbf{r}') \right.$$
$$\left. + [\mathbf{e}_n \times \mathbf{U}(\mathbf{r})] \cdot [\nabla \times \mathcal{G}(\mathbf{r}, \mathbf{r}')] \right\}. \qquad (17\text{-}4)$$

But the integral on the infinitely large and faraway surface S_∞ goes to zero since the Green's dyadic decays with increasing distance; furthermore,

$$e_n \cdot [U(r) \times \mathfrak{G}(r, r')] = [e_n \times U(r)] \cdot \mathfrak{G}(r, r'). \qquad (17\text{-}5)$$

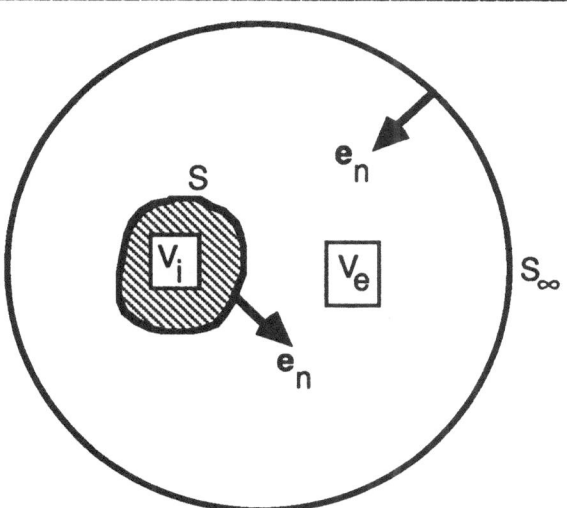

Fig. 17-1. Relevant to Huygens's principle for isotropic chiral media.

Finally, on also utilizing the transpose properties (13-15) of the Green's dyadic, (17-3) simplifies to

$$\int_{V_e} dv \left\{ U(r) \cdot \mathbf{I}\delta(r - r') \right\} = -2\gamma^2\beta \int_S ds \, \mathfrak{G}(r', r) \cdot [e_n \times U(r)]$$
$$+ \int_S ds \left\{ \mathfrak{G}(r', r) \cdot \left(e_n \times [\nabla \times U(r)] \right) \right.$$
$$\left. + [\nabla' \times \mathfrak{G}(r', r)] \cdot \left(e_n \times U(r) \right) \right\}. \qquad (17\text{-}6)$$

In addition, the Dirac delta function can be used to eliminate the integral on the left hand side of (17-6), and Huygens's principle for chiral media can be simply stated *via* the twin relations

$$U(r') = -2\gamma^2\beta \int_S ds \, \mathfrak{G}(r', r) \cdot [e_n \times U(r)]$$
$$+ \int_S ds \left\{ \mathfrak{G}(r', r) \cdot \left(e_n \times [\nabla \times U(r)] \right) \right.$$
$$\left. + [\nabla' \times \mathfrak{G}(r', r)] \cdot \left(e_n \times U(r) \right) \right\}; \, r' \in V_e, \qquad (17\text{-}7a)$$

$$0 = -2\gamma^2\beta\int_S ds \; \circledB(\mathbf{r}', \mathbf{r})\cdot[\mathbf{e}_n\times\mathbf{U}(\mathbf{r})]$$
$$+\int_S ds \; \{ \circledB(\mathbf{r}', \mathbf{r})\cdot(\mathbf{e}_n\times[\nabla\times\mathbf{U}(\mathbf{r})])$$
$$+ [\nabla'\times\circledB(\mathbf{r}', \mathbf{r})]\cdot(\mathbf{e}_n\times\mathbf{U}(\mathbf{r})) \}; \; \mathbf{r}' \in V_i. \qquad (17\text{-}7b)$$

Thus, it is possible to compute the field $\mathbf{U}$ in any source-free region provided the tangential components of $\mathbf{U}$ and $\nabla\times\mathbf{U}(\mathbf{r})$ are known on the surfaces enclosing the volume of interest.

Parenthetically, let (17-7) specialize to both $\mathbf{U} = \mathbf{E}$ and $\mathbf{U} = \mathbf{H}$. Then, by using (4-2) to get rid of the $\nabla\times\mathbf{U}$ terms, and by utilizing the inverse Bohren decomposition (15-3), Huygens's principle for the LCP field $\mathbf{Q}_1$ can be obtained as

$$\mathbf{Q}_1(\mathbf{r}') = (2\gamma^2/k)\int_S ds \; \circledB_1(\mathbf{r}', \mathbf{r})\cdot[\mathbf{e}_n\times\mathbf{Q}_1(\mathbf{r})]; \; \mathbf{r}' \in V_e, \qquad (17\text{-}8a)$$
$$0 = (2\gamma^2/k)\int_S ds \; \circledB_1(\mathbf{r}', \mathbf{r})\cdot[\mathbf{e}_n\times\mathbf{Q}_1(\mathbf{r})]; \; \mathbf{r}' \in V_i. \qquad (17\text{-}8b)$$

In a similar manner, Huygens's principle for the RCP field $\mathbf{Q}_2$ is given as

$$\mathbf{Q}_2(\mathbf{r}') = - (2\gamma^2/k)\int_S ds \; \circledB_2(\mathbf{r}', \mathbf{r})\cdot[\mathbf{e}_n\times\mathbf{Q}_2(\mathbf{r})]; \; \mathbf{r}' \in V_e, \qquad (17\text{-}9a)$$
$$0 = - (2\gamma^2/k)\int_S ds \; \circledB_2(\mathbf{r}', \mathbf{r})\cdot[\mathbf{e}_n\times\mathbf{Q}_2(\mathbf{r})]; \; \mathbf{r}' \in V_i. \qquad (17\text{-}9b)$$

Thus, the independence of the LCP and the RCP fields in isotropic chiral media is reaffirmed by (17-8) and (17-9); it should not be forgotten, however, that boundary conditions on S are prescribed on the tangential components of $\mathbf{E}$ and $\mathbf{H}$.

The development of Huygens's principle contains the seeds of a scattering formalism pertinent to isotropic chiral media. Let V_i denote a scatterer on which a field $\mathbf{U}^{inc}$ is incident. As per the equivalence principle [Barber & Yeh 1975], the incident field will cause the creation of surface currents, $\mathbf{e}_n\times[\nabla_+\times\mathbf{U}(\mathbf{r})]$ and $\mathbf{e}_n\times\mathbf{U}_+(\mathbf{r})$, on S_+, the exterior surface of the obstacle. If V_i were to be perfectly conducting and $\mathbf{U}$ were to represent either $\mathbf{E}$ or $\mathbf{H}$, $\mathbf{e}_n\times\mathbf{E}_+$ would be identically zero on S. On the other hand, if V_i were to be electromagnetically permeable, then these currents would be merely *equivalent* surface currents [Barber & Yeh 1975]. In either case, the effect of these surface currents would be to null the incident field inside the obstacle. Hence, from (17-7b)

$$-\mathbf{U}^{inc}(\mathbf{r}') = -2\gamma^2\beta\!\int_S \; ds \; \mathcal{G}(\mathbf{r}', \mathbf{r})\boldsymbol{\cdot}[\mathbf{e}_n\times\mathbf{U}_+(\mathbf{r})]$$
$$+\!\int_S \; ds \; \big\{ \mathcal{G}(\mathbf{r}', \mathbf{r})\boldsymbol{\cdot}\big(\mathbf{e}_n \times [\nabla_+\times\mathbf{U}(\mathbf{r})]\big)$$
$$+ [\nabla'\times\mathcal{G}(\mathbf{r}', \mathbf{r})]\boldsymbol{\cdot}\big(\mathbf{e}_n\times\mathbf{U}_+(\mathbf{r})\big) \big\}, \mathbf{r}' \in V_i \qquad (17\text{-}10)$$

constitutes an integral equation for the surface fields and, as such, it is the extinction theorem [Waterman 1969] for scattering in isotropic chiral media. Once it is solved, the scattered field can be simply computed through (17-7a) as

$$\mathbf{U}^{sc}(\mathbf{r}') = -2\gamma^2\beta\!\int_S \; ds \; \mathcal{G}(\mathbf{r}', \mathbf{r})\boldsymbol{\cdot}[\mathbf{e}_n\times\mathbf{U}_+(\mathbf{r})]$$
$$+\!\int_S \; ds \; \big\{ \mathcal{G}(\mathbf{r}', \mathbf{r})\boldsymbol{\cdot}\big(\mathbf{e}_n \times [\nabla_+\times\mathbf{U}(\mathbf{r})]\big)$$
$$+ [\nabla'\times\mathcal{G}(\mathbf{r}', \mathbf{r})]\boldsymbol{\cdot}\big(\mathbf{e}_n\times\mathbf{U}_+(\mathbf{r})\big) \big\}, \mathbf{r}' \in V_e. \qquad (17\text{-}11)$$

The properties of the scatterer need also to be taken into account. If V_i were to be impenetrable, then the reductions of (17-10) and (17-11) are self-evident. For a permeable V_i, on the other hand, let the discussion specialize to $\mathbf{U} = \mathbf{E}$ or $\mathbf{H}$. In which case, the boundary conditions are such that S remains charge- as well as current-neutral. Consequently,

$$\mathbf{e}_n\times[\nabla_+\times\mathbf{U}(\mathbf{r})] = \mathbf{e}_n\times[\nabla\times\mathbf{U}^{int}(\mathbf{r})]; \quad \mathbf{r} \in S, \qquad (17\text{-}12a)$$
$$\mathbf{e}_n\times\mathbf{U}_+(\mathbf{r}) = \mathbf{e}_n\times\mathbf{U}^{int}(\mathbf{r}); \quad \mathbf{r} \in S, \qquad (17\text{-}12b)$$

should be substituted in (17-10) and (17-11) to yield, respectively,

$$-\mathbf{U}^{inc}(\mathbf{r}') = -2\gamma^2\beta\!\int_S \; ds \; \mathcal{G}(\mathbf{r}', \mathbf{r})\boldsymbol{\cdot}[\mathbf{e}_n\times\mathbf{U}^{int}(\mathbf{r})]$$
$$+\!\int_S \; ds \; \big\{ \mathcal{G}(\mathbf{r}', \mathbf{r})\boldsymbol{\cdot}\big(\mathbf{e}_n \times [\nabla\times\mathbf{U}^{int}(\mathbf{r})]\big)$$
$$+ [\nabla'\times\mathcal{G}(\mathbf{r}', \mathbf{r})]\boldsymbol{\cdot}\big(\mathbf{e}_n\times\mathbf{U}^{int}(\mathbf{r})\big) \big\}, \mathbf{r}' \in V_i, \qquad (17\text{-}13)$$

$$\mathbf{U}^{sc}(\mathbf{r}') = -2\gamma^2\beta\!\int_S \; ds \; \mathcal{G}(\mathbf{r}', \mathbf{r})\boldsymbol{\cdot}[\mathbf{e}_n\times\mathbf{U}^{int}(\mathbf{r})]$$
$$+\!\int_S \; ds \; \big\{ \mathcal{G}(\mathbf{r}', \mathbf{r})\boldsymbol{\cdot}\big(\mathbf{e}_n \times [\nabla\times\mathbf{U}^{int}(\mathbf{r})]\big)$$
$$+ [\nabla'\times\mathcal{G}(\mathbf{r}', \mathbf{r})]\boldsymbol{\cdot}\big(\mathbf{e}_n\times\mathbf{U}^{int}(\mathbf{r})\big) \big\}, \mathbf{r}' \in V_e. \qquad (17\text{-}14)$$

The resulting sets of equations, (17-13) and (17-14), can be most easily solved, for example, by the T-matrix procedure [Waterman 1969; Varadan & Varadan

1980]. It is to be noted that only the Green's function of the exterior region is utilized here. The formalism is largely independent of the constitution of the interior region V_i, which enters into the picture solely through the representations for $\mathbf{U}^{int}$ and $\nabla\times\mathbf{U}^{int}$ in the boundary conditions (17-12).

Equations (17-13) and (17-14) shed some light on the nature of the scattered field, given the incident field. On specializing these equations to $\mathbf{U} = \mathbf{E}$ and $\mathbf{U} = \mathbf{H}$, and utilizing the inverse Bohren decomposition (15-3), it is possible to obtain

$$-\mathbf{Q}_1^{inc}(\mathbf{r}') = \gamma^2 k^{-1}\int_S ds\ \boldsymbol{\mathfrak{G}}_1(\mathbf{r}', \mathbf{r})\cdot\mathbf{e}_n\times[\mathbf{E}^{int}(\mathbf{r})-a_R\mathbf{H}^{int}(\mathbf{r})],\quad \mathbf{r}'\in V_i,\ (17\text{-}15a)$$

$$\mathbf{Q}_2^{inc}(\mathbf{r}') = \gamma^2 k^{-1}\int_S ds\ \boldsymbol{\mathfrak{G}}_2(\mathbf{r}', \mathbf{r})\cdot\mathbf{e}_n\times[\mathbf{H}^{int}(\mathbf{r})-a_L\mathbf{E}^{int}(\mathbf{r})],\quad \mathbf{r}'\in V_i,\ (17\text{-}15b)$$

$$\mathbf{Q}_1^{sc}(\mathbf{r}') = \gamma^2 k^{-1}\int_S ds\ \boldsymbol{\mathfrak{G}}_1(\mathbf{r}', \mathbf{r})\cdot\mathbf{e}_n\times[\mathbf{E}^{int}(\mathbf{r})-a_R\mathbf{H}^{int}(\mathbf{r})],\quad \mathbf{r}'\in V_e,\ (17\text{-}15c)$$

$$-\mathbf{Q}_2^{sc}(\mathbf{r}') = \gamma^2 k^{-1}\int_S ds\ \boldsymbol{\mathfrak{G}}_2(\mathbf{r}', \mathbf{r})\cdot\mathbf{e}_n\times[\mathbf{H}^{int}(\mathbf{r})-a_L\mathbf{E}^{int}(\mathbf{r})],\quad \mathbf{r}'\in V_e,\ (17\text{-}15d)$$

Let the scatterer V_i be also chiral with constitutive parameters ε_i, μ_i, and β_i; the quantities $a_{Ri} = -i\sqrt{[\mu_i/\varepsilon_i]} = -1/a_{Li}$. The decomposition (15-30 is equally applicable to the scatterer as well; consequently, the foregoing set of integral equations is transformed to

$$-\mathbf{Q}_1^{inc}(\mathbf{r}') = \gamma^2 k^{-1}[1+a_R/a_{Ri}]\int_S ds\ \boldsymbol{\mathfrak{G}}_1(\mathbf{r}', \mathbf{r})\cdot[\mathbf{e}_n\times \mathbf{Q}_1^{int}(\mathbf{r})]$$
$$+ \gamma^2 k^{-1}[a_{Ri}-a_R]\int_S ds\ \boldsymbol{\mathfrak{G}}_1(\mathbf{r}', \mathbf{r})\cdot[\mathbf{e}_n\times \mathbf{Q}_2^{int}(\mathbf{r})],\quad \mathbf{r}'\in V_i,\ (17\text{-}16a)$$

$$\mathbf{Q}_1^{sc}(\mathbf{r}') = \gamma^2 k^{-1}[1+a_R/a_{Ri}]\int_S ds\ \boldsymbol{\mathfrak{G}}_1(\mathbf{r}', \mathbf{r})\cdot[\mathbf{e}_n\times \mathbf{Q}_1^{int}(\mathbf{r})]$$
$$+ \gamma^2 k^{-1}[a_{Ri}-a_R]\int_S ds\ \boldsymbol{\mathfrak{G}}_1(\mathbf{r}', \mathbf{r})\cdot[\mathbf{e}_n\times \mathbf{Q}_2^{int}(\mathbf{r})],\quad \mathbf{r}'\in V_e,\ (17\text{-}16b)$$

$$-\mathbf{Q}_2^{inc}(\mathbf{r}') = \gamma^2 k^{-1}[a_L-a_{Li}]\int_S ds\ \boldsymbol{\mathfrak{G}}_2(\mathbf{r}', \mathbf{r})\cdot[\mathbf{e}_n\times \mathbf{Q}_1^{int}(\mathbf{r})]$$
$$- \gamma^2 k^{-1}[1+a_{Ri}/a_R]\int_S ds\ \boldsymbol{\mathfrak{G}}_2(\mathbf{r}', \mathbf{r})\cdot[\mathbf{e}_n\times \mathbf{Q}_2^{int}(\mathbf{r})],\quad \mathbf{r}'\in V_i,\ (17\text{-}16c)$$

$$\mathbf{Q}_2^{sc}(\mathbf{r}') = \gamma^2 k^{-1}[a_L-a_{Li}]\int_S ds\ \boldsymbol{\mathfrak{G}}_2(\mathbf{r}', \mathbf{r})\cdot[\mathbf{e}_n\times \mathbf{Q}_1^{int}(\mathbf{r})]$$
$$- \gamma^2 k^{-1}[1+a_{Ri}/a_R]\int_S ds\ \boldsymbol{\mathfrak{G}}_2(\mathbf{r}', \mathbf{r})\cdot[\mathbf{e}_n\times \mathbf{Q}_2^{int}(\mathbf{r})],\quad \mathbf{r}'\in V_e.\ (17\text{-}16d)$$

Quite remarkably, (17-16) show that both the scattered and the internally induced fields contain left- as well as right- circularly polarized components regardless of the state of polarization of the incident field. One notable case is worthy of mention: let the scatterer be impedance-matched with the surrounding medium; i.e., let $\mu_i/\varepsilon_i = \mu/\varepsilon$. In that case, (17-16) simplify to

$$-Q_1^{inc}(\mathbf{r}') = 2\gamma^2 k^{-1} \int_S ds\ \mathfrak{B}_1(\mathbf{r}', \mathbf{r}) \cdot [\mathbf{e}_n \times Q_1^{int}(\mathbf{r})], \qquad \mathbf{r}' \in V_i, \qquad (17\text{-}17a)$$

$$Q_1^{sc}(\mathbf{r}') = 2\gamma^2 k^{-1} \int_S ds\ \mathfrak{B}_1(\mathbf{r}', \mathbf{r}) \cdot [\mathbf{e}_n \times Q_1^{int}(\mathbf{r})], \qquad \mathbf{r}' \in V_e, \qquad (17\text{-}17b)$$

and

$$Q_2^{inc}(\mathbf{r}') = 2\gamma^2 k^{-1} \int_S ds\ \mathfrak{B}_2(\mathbf{r}', \mathbf{r}) \cdot [\mathbf{e}_n \times Q_2^{int}(\mathbf{r})], \qquad \mathbf{r}' \in V_i, \qquad (17\text{-}18a)$$

$$-Q_2^{sc}(\mathbf{r}') = 2\gamma^2 k^{-1} \int_S ds\ \mathfrak{B}_2(\mathbf{r}', \mathbf{r}) \cdot [\mathbf{e}_n \times Q_2^{int}(\mathbf{r})], \qquad \mathbf{r}' \in V_e. \qquad (17\text{-}18b)$$

Thus, if the scatterer and the surrounding medium are impedance-matched, the scattered as well as the internal fields have the same state of polarization as the incident field. This conclusion can be verified by a more elementary consideration: the tangential components of the **E** and the **H** fields must be continuous across S. Consequently,

$$\mathbf{e}_n \times [Q_1^{inc}(\mathbf{r}) + a_R Q_2^{inc}(\mathbf{r})] + \mathbf{e}_n \times [Q_1^{sc}(\mathbf{r}) + a_R Q_2^{sc}(\mathbf{r})] =$$
$$\mathbf{e}_n \times [Q_1^{int}(\mathbf{r}) + a_{Ri} Q_2^{int}(\mathbf{r})], \qquad \mathbf{r} \in S, \qquad (17\text{-}19a)$$

$$\mathbf{e}_n \times [a_L Q_1^{inc}(\mathbf{r}) + Q_2^{inc}(\mathbf{r})] + \mathbf{e}_n \times [a_L Q_1^{sc}(\mathbf{r}) + Q_2^{sc}(\mathbf{r})] =$$
$$\mathbf{e}_n \times [a_{Li} Q_1^{int}(\mathbf{r}) + Q_2^{int}(\mathbf{r})], \qquad \mathbf{r} \in S. \qquad (17\text{-}19b)$$

On enforcing the condition $\mu_i/\varepsilon_i = \mu/\varepsilon$ in (17-19a,b), it is easy to see that the boundary conditions simplify to

$$\mathbf{e}_n \times [Q_1^{inc}(\mathbf{r}) + Q_1^{sc}(\mathbf{r})] = \mathbf{e}_n \times Q_1^{int}(\mathbf{r}), \qquad \mathbf{r} \in S, \qquad (17\text{-}20a)$$

$$\mathbf{e}_n \times [Q_2^{inc}(\mathbf{r}) + Q_2^{sc}(\mathbf{r})] = \mathbf{e}_n \times Q_2^{int}(\mathbf{r}), \qquad \mathbf{r} \in S, \qquad (17\text{-}20b)$$

proving that the conclusion made in this paragraph is indeed correct.

18. PLANE WAVE SCATTERING IN CHIRAL MEDIA

Since Huygens's principle (17-7a,b) applies to any field which does not have sources in V_e in Fig. 18-1, it can be used to obtain a scattering matrix which characterizes the frequency response of the obstacle volume V_i. It is, however, easier to consider the LCP and the RCP components of the scattered field and use the dyadics (13-17b,c).

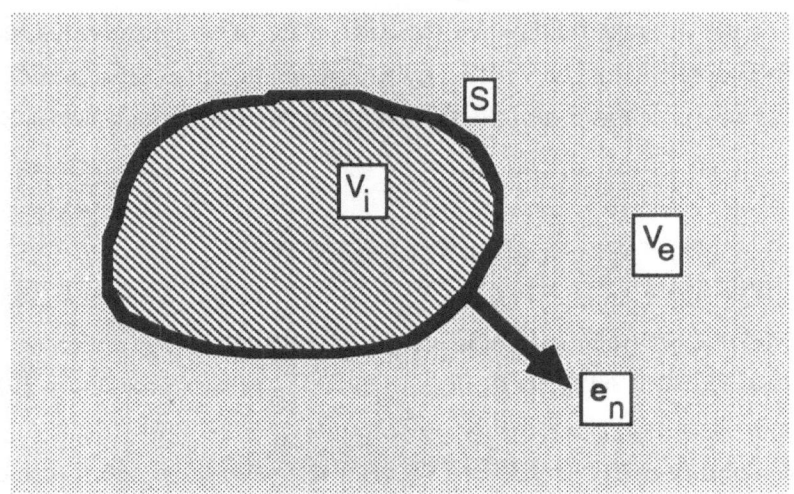

Fig. 18-1 For the far-zone scattering amplitude

On applying in V_e, the Green's second identity for vector $\mathbf{P}$ and dyadic $\boldsymbol{\mathfrak{P}}$ derived in Section 17, one obtains

$$\int_{V_e} dv \left\{ \mathbf{P} \cdot (\nabla \times \nabla \times \boldsymbol{\mathfrak{P}}) - (\nabla \times \nabla \times \mathbf{P}) \cdot \boldsymbol{\mathfrak{P}} \right\}$$
$$= \int_S ds \left\{ \mathbf{e}_n \times (\nabla \times \mathbf{P}) \cdot \boldsymbol{\mathfrak{P}} + (\mathbf{e}_n \times \mathbf{P}) \cdot (\nabla \times \boldsymbol{\mathfrak{P}}) \right\}, \qquad (18-1)$$

with reference to Figure 18-1; it is assumed here that $\boldsymbol{\mathfrak{P}}$ satisfies the radiation conditions at infinity. First of all, in (17-2) let $\mathbf{P} = \mathbf{Q}_1{}^{sc}$, the LCP component of the scattered field; and let $\boldsymbol{\mathfrak{P}} = \Gamma_1$ of (13-17b). Then, the utilization of the relations

$$\nabla \times \nabla \times \Gamma_1(\mathbf{r},\mathbf{r}') - \gamma_1^2 \, \Gamma_1(\mathbf{r},\mathbf{r}') = \mathbf{I}\delta(\mathbf{r}\text{-}\mathbf{r}'), \tag{18-2a}$$

$$\nabla \times \nabla \times \mathbf{Q}_1{}^{sc}(\mathbf{r}) - \gamma_1^2 \, \mathbf{Q}_1{}^{sc}(\mathbf{r}) = \mathbf{0}, \, \mathbf{r} \in V_e, \tag{18-2b}$$

leads to the integral equation

$$\mathbf{Q}_1{}^{sc}(\mathbf{r}') = 2\gamma_1 \int_S \, ds \, \Gamma_1(\mathbf{r}',\mathbf{r}) \cdot \left(\mathbf{e}_n \times \mathbf{Q}_1{}^{sc}(\mathbf{r}) \right), \, \mathbf{r}' \in V_e. \tag{18-3a}$$

Next, the derivation of the relation

$$\mathbf{Q}_2{}^{sc}(\mathbf{r}') = -2\gamma_2 \int_S \, ds \, \Gamma_2(\mathbf{r}',\mathbf{r}) \cdot \left(\mathbf{e}_n \times \mathbf{Q}_2{}^{sc}(\mathbf{r}) \right), \, \mathbf{r}' \in V_e. \tag{18-3b}$$

follows a similar pattern as well. On applying the inverse Bohren decomposition (15-3) to the equations (18-3a,b), it is easy to show that

$$\begin{aligned}
\mathbf{E}^{sc}(\mathbf{r}') = \int_S \, ds \, &\left\{ \gamma_1 \Gamma_1(\mathbf{r}',\mathbf{r}) - \gamma_2 \Gamma_2(\mathbf{r}',\mathbf{r}) \right\} \cdot \left(\mathbf{e}_n \times \mathbf{E}^{sc}(\mathbf{r}) \right) \\
&+ (i\omega\mu/k) \int_S \, ds \, \left\{ \gamma_1 \Gamma_1(\mathbf{r}',\mathbf{r}) + \gamma_2 \Gamma_2(\mathbf{r}',\mathbf{r}) \right\} \\
&\qquad\qquad\qquad \cdot \left(\mathbf{e}_n \times \mathbf{H}^{sc}(\mathbf{r}) \right), \, \mathbf{r}' \in V_e,
\end{aligned} \tag{18-4a}$$

$$\begin{aligned}
\mathbf{H}^{sc}(\mathbf{r}') = \int_S \, ds \, &\left\{ \gamma_1 \Gamma_1(\mathbf{r}',\mathbf{r}) - \gamma_2 \Gamma_2(\mathbf{r}',\mathbf{r}) \right\} \cdot \left(\mathbf{e}_n \times \mathbf{H}^{sc}(\mathbf{r}) \right) \\
&- (i\omega\varepsilon/k) \int_S \, ds \, \left\{ \gamma_1 \Gamma_1(\mathbf{r}',\mathbf{r}) + \gamma_2 \Gamma_2(\mathbf{r}',\mathbf{r}) \right\} \\
&\qquad\qquad\qquad \cdot \left(\mathbf{e}_n \times \mathbf{E}^{sc}(\mathbf{r}) \right), \, \mathbf{r}' \in V_e.
\end{aligned} \tag{18-4b}$$

Now, in the far zone the scattered field must be completely TEM; and it will consist, in general, of both LCP and RCP waves as is evident from the derivation of (18-4a,b). To evaluate the radiation field in a direction $\mathbf{e}_\eta$, the asymptotic forms of Γ_1 and Γ_2 are to be used:

$$\Gamma_1(\mathbf{r}', \mathbf{r}) \approx \mathbf{I} \, (1/4\pi) \, \{\exp[i\gamma_1 r']/r'\} \, \exp[-i\gamma_1 \, \mathbf{e}_\eta \cdot \mathbf{r}], \tag{18-5a}$$

$$\Gamma_2(\mathbf{r}', \mathbf{r}) \approx \mathbf{I} \, (1/4\pi) \, \{\exp[i\gamma_2 r']/r'\} \, \exp[-i\gamma_2 \, \mathbf{e}_\eta \cdot \mathbf{r}], \tag{18-5b}$$

$\mathbf{e}_\eta$ being a vector of unit magnitude. Substitution of (18-5a,b) into (18-4a,b) gives rise to the simple relationship,

$$E^{sc}(r') = \{exp[i\gamma_1 r']/r'\} \; F_{t1}(e_\eta) + \{exp[i\gamma_2 r']/r'\} \; F_{t2}(e_\eta), \quad (18\text{-}6a)$$

$$(i\omega\mu/k)H^{sc}(r') = \{exp[i\gamma_1 r']/r'\} \; F_{t1}(e_\eta) - \{exp[i\gamma_2 r']/r'\} \; F_{t2}(e_\eta), \quad (18\text{-}6b)$$

where,

$$F_{t1}(e_\eta) = (ik\gamma_1/4\omega\epsilon\pi)\!\int_S ds \; e_n \times$$
$$\{H^{sc}(r) - (i\omega\epsilon/k)E^{sc}(r)\}exp[-i\gamma_1 e_\eta \cdot r], \quad (18\text{-}7a)$$

$$F_{t2}(e_\eta) = (ik\gamma_2/4\omega\epsilon\pi)\!\int_S ds \; e_n \times$$
$$\{H^{sc}(r) + (i\omega\epsilon/k)E^{sc}(r)\}exp[-i\gamma_2 e_\eta \cdot r], \quad (18\text{-}7b)$$

are the far-zone amplitudes of the LCP and the RCP scattered fields, respectively, expressed as functions of the scattered field on the surface S of the obstacle.

The scattered fields on the surface are nothing but parts of the surface current densities, $e_n \times H_+$ and $e_n \times E_+$, excited due to the irradiation of V_i by an incoming field. Let the incident field be either a LCP planewave,

$$E_1^{inc}(r) = e_1 \; exp \; [i\gamma_1 \; e_{\eta 1} \cdot r], \quad (18\text{-}8a)$$

$$H_1^{inc}(r) = -(i\omega\epsilon/k)e_1 \; exp \; [i\gamma_1 \; e_{\eta 1} \cdot r], \quad (18\text{-}8b)$$

$$e_1 \cdot e_{\eta 1} = 0, \quad (18\text{-}8c)$$

$$e_{\eta 1} \times e_1 = -ie_1, \quad (18\text{-}8d)$$

or a RCP planewave,

$$E_2^{inc}(r) = e_2 \; exp \; [i\gamma_2 \; e_{\eta 2} \cdot r], \quad (18\text{-}9a)$$

$$H_2^{inc}(r) = (i\omega\epsilon/k)e_2 \; exp \; [i\gamma_2 \; e_{\eta 2} \cdot r], \quad (18\text{-}9b)$$

$$e_2 \cdot e_{\eta 2} = 0, \quad (18\text{-}9c)$$

$$e_{\eta 2} \times e_2 = ie_2, \quad (18\text{-}9d)$$

or a combination of both. From the far-zone point of view then, planewave scattering matrices $\mathbf{S}_{nm}$, (n,m = 1,2) can be construed [van Bladel 1985] in the following fashion:

$$F_{t1}(e_\eta) = \mathbf{S}_{11}(e_\eta|e_{\eta 1}) \cdot e_1 + \mathbf{S}_{12}(e_\eta|e_{\eta 2}) \cdot e_2, \quad (18\text{-}10a)$$

$$F_{t2}(e_\eta) = \mathbf{S}_{21}(e_\eta|e_{\eta 1}) \cdot e_1 + \mathbf{S}_{22}(e_\eta|e_{\eta 2}) \cdot e_2. \tag{18-10b}$$

An interesting property of these scattering matrices comes from the reciprocal nature of the chiral medium. Let $\{E, H\}$ be the *total* fields existing in V_e due to any linear combination of the two incident fields $\{E_m^{inc}, H_m^{inc}\}|_{m=1,2}$ as in (18-8) and (18-9); let $\{E'', H''\}$ be the *total* fields existing in V_e independently due to the sum of the two incident fields $\{E''_m{}^{inc}, H''_m{}^{inc}\}|_{m=1,2}$ which resemble (18-8) and (18-9). Realizing that the "equivalent" currents induced on the exterior surface of the obstacle are the sources of the scattered field in the exterior region, the reaction theorem (5-5) can be used to yield

$$\Sigma_{m=1,2} \int_S ds \left[\{e_n \times H^{sc}(r)\} \cdot E''_m{}^{inc}(r) + \{e_n \times E^{sc}(r)\} \cdot H''_m{}^{inc}(r) \right]$$
$$= \Sigma_{m=1,2} \int_S ds \left[\{e_n \times H''^{sc}(r)\} \cdot E_m^{inc}(r) + \right.$$
$$\left. \{e_n \times E''^{sc}(r)\} \cdot H_m^{inc}(r) \right]. \tag{18-11}$$

Substitution of the explicit expressions (18-8) and (18-9) of the incident fields and the use of (18-10a,b) transforms (18-11) into the equation

$$\left[e_1'' \cdot \mathbf{S}_{11}(-e_{\eta 1}''|e_{\eta 1}) \cdot e_1 - e_1 \cdot \mathbf{S}_{11}(-e_{\eta 1}|e_{\eta 1}'') \cdot e_1'' \right] +$$
$$\left[e_1'' \cdot \mathbf{S}_{12}(-e_{\eta 1}''|e_{\eta 2}) \cdot e_2 - e_2 \cdot \mathbf{S}_{21}(-e_{\eta 2}|e_{\eta 1}'') \cdot e_1'' \right] +$$
$$\left[e_2'' \cdot \mathbf{S}_{21}(-e_{\eta 2}''|e_{\eta 1}) \cdot e_1 - e_1 \cdot \mathbf{S}_{12}(-e_{\eta 1}|e_{\eta 2}'') \cdot e_2'' \right] +$$
$$\left[e_2'' \cdot \mathbf{S}_{22}(-e_{\eta 2}''|e_{\eta 2}) \cdot e_2 - e_1 \cdot \mathbf{S}_{22}(-e_{\eta 2}|e_{\eta 2}'') \cdot e_2'' \right] = 0, \tag{18-12}$$

which implies that

$$\mathbf{S}_{11}(-e_{\eta 1}''|e_{\eta 1}) = [\mathbf{S}_{11}(-e_{\eta 1}|e_{\eta 1}'')]^{(tr)}, \tag{18-13a}$$
$$\mathbf{S}_{22}(-e_{\eta 2}''|e_{\eta 2}) = [\mathbf{S}_{22}(-e_{\eta 2}|e_{\eta 2}'')]^{(tr)}, \tag{18-13b}$$
$$\mathbf{S}_{12}(-e_{\eta 1}''|e_{\eta 2}) = [\mathbf{S}_{21}(-e_{\eta 2}|e_{\eta 1}'')]^{(tr)}, \tag{18-13c}$$
$$\mathbf{S}_{21}(-e_{\eta 2}''|e_{\eta 1}) = [\mathbf{S}_{12}(-e_{\eta 1}|e_{\eta 2}'')]^{(tr)}. \tag{18-13d}$$

The essence of this argument is that the supermatrix $\mathscr{S}$, given by

$$\mathscr{S} = \begin{bmatrix} \mathscr{S}_{11} & \mathscr{S}_{12} \\ \mathscr{S}_{21} & \mathscr{S}_{22} \end{bmatrix}, \tag{18-14}$$

is symmetric, and the scattering formalism of (17-11) and (17-12) satisfies reciprocity constraints imposed upon it by the self-complementary nature of the medium filling the external volume V_e, regardless of the the kind of medium which constitutes the scattering object V_i. It should be noted that $\mathscr{S}_{12} = \mathscr{S}_{21} = 0$ and $\mathscr{S}_{11} = \mathscr{S}_{22}$ are symmetric matrices if $\beta = 0$, and conform to the result due to de Hoop [1960] for isotropic achiral media.

As in the case of the isotropic achiral media, here too the forward planewave-scattering amplitude contains information about the loss suffered by an incident planewave in the forward direction. Now, the time-averaged power *scattered* by the obstacle volume V_i is given by

$$P_{sca} = (1/2) \, Re \left\{ \int_S ds \, e_n \cdot E^{sc} \times H^{sc*} \right\}, \tag{18-15}$$

while the time-averaged power *absorbed* in the scatterer is

$$P_{abs} = -(1/2) \, Re \left\{ \int_S ds \, e_n \cdot [E^{sc}+E^{inc}] \times [H^{sc*}+H^{inc*}] \right\}. \tag{18-16}$$

Since the average power carried across S by the incident wave has to be zero, it can, consequently, be shown that the *total* power extracted from the incident wave is given by

$$P_{ext} = -(1/2) \, Re \left\{ \int_S ds \, e_n \cdot [E^{inc} \times H^{sc*} + E^{sc} \times H^{inc*}] \right\}. \tag{18-17}$$

First of all, let the scatterer be irradiated by the LCP planewave (18-8). From (18-17), it turns out that the total power extinguished by the obstacle is given by

$$P_{ext1} = (1/2)Re\left(\mathbf{e}_1{}^* \cdot \int_S d\mathbf{s}\ \mathbf{e}_n \times\right.$$
$$\left.\{\mathbf{H}^{sc}(\mathbf{r}) - (i\omega\varepsilon/k)\mathbf{E}^{sc}(\mathbf{r})\}\exp[-i\gamma_1\ \mathbf{e}_{\eta1}\cdot\mathbf{r}]\right), \qquad (18\text{-}18)$$

while the extinction cross-section is defined by,

$$C_{ext1} = (2k/\omega\varepsilon)\ |\mathbf{e}_1|^{-2}\ P_{ext1}. \qquad (18\text{-}19)$$

Comparison of (18-7a) and (18-18) leads to the forward amplitude theorem for LCP planewave incidence:

$$C_{ext1} = (4\pi/\gamma_1)\ Im\left(\mathbf{e}_1{}^* \cdot \mathbf{F}_{t1}(\mathbf{e}_{\eta1})\ |\mathbf{e}_1|^{-2}\right). \qquad (18\text{-}20)$$

Likewise, for RCP planewave incidence of the type (18-9), the total power extracted can be set down as

$$P_{ext2} = (1/2)Re\left(\mathbf{e}_2{}^* \cdot \int_S d\mathbf{s}\ \mathbf{e}_n \times\right.$$
$$\left.\{\mathbf{H}^{sc}(\mathbf{r}) + (i\omega\varepsilon/k)\mathbf{E}^{sc}(\mathbf{r})\}\exp[-i\gamma_2\ \mathbf{e}_{\eta2}\cdot\mathbf{r}]\right), \qquad (18\text{-}21)$$

whence the forward amplitude theorem for RCP planewave incidence,

$$C_{ext2} = (4\pi/\gamma_2)\ Im\left(\mathbf{e}_2{}^* \cdot \mathbf{F}_{t2}(\mathbf{e}_{\eta2})\ |\mathbf{e}_2|^{-2}\right). \qquad (18\text{-}22)$$

The similarity in form of (18-20) and (18-22) on one hand, and of either with the forward amplitude theorem [de Hoop 1959] for linearly polarized waves in isotropic achiral media, should be noted and would be of value in verifying scattering calculations. It is to be emphasized that (18-20) and (18-22) are exact relations, and if approximate values of $\mathbf{F}_{t1}$ and $\mathbf{F}_{t2}$ are used then *manifestly wrong* [Jackson 1975] results may be obtained.

19. A SCALAR TREATMENT

In the previous sections, the electromagnetic theory for chiral media has received a vector treatment and has involved dyadics. A purely scalar treatment is also possible and has been given by Weiglhofer [1988] based on the scalarisation of Maxwell's equations [Weiglhofer 1987]. This procedure is based upon the prescription of an arbitrary unit vector e_b, and the subsequent decomposition of any vector field in terms of a component parallel to e_b and another one transverse to it. In other words, a vector field A is decomposed into

$$A = A_t + A_b\, e_b; \qquad\qquad e_b\cdot A_t = 0. \qquad\qquad (19\text{-}1)$$

Consequently, (4-9a,b) can be rewritten, respectively as

$$E = -(i\omega\varepsilon)^{-1}[\nabla_t\cdot(H_t + i\omega\varepsilon\beta E_t) - J], \qquad\qquad (19\text{-}2a)$$
$$H = (i\omega\mu)^{-1}[\nabla_t\cdot(E_t - i\omega\mu\beta H_t) + K], \qquad\qquad (19\text{-}2b)$$

in which equations,

$$\nabla = \nabla_t + (\partial/\partial x_b)\, e_b. \qquad\qquad (19\text{-}3)$$

The solution of (19-2) can then be found out in terms of scalar Hertz potentials u and v as

$$E = \nabla\times\nabla\times ue_b + \gamma^2\beta\nabla\times ue_b + i\omega\mu(\gamma/k)^2\nabla\times ve_b$$
$$+ (i\omega\varepsilon)^{-1} J_b e_b + \nabla_t q_1, \qquad\qquad (19\text{-}4a)$$
$$H = \nabla\times\nabla\times ve_b + \gamma^2\beta\nabla\times ve_b - i\omega\varepsilon(\gamma/k)^2\nabla\times ue_b$$
$$+ (i\omega\mu)^{-1} K_b e_b + \nabla_t q_2, \qquad\qquad (19\text{-}4b)$$

if the scalar potentials themselves satisfy the relations

$$[\nabla^2 + \gamma^4 k^{-2}(1+k^2\beta^2)]\, u\ + 2i\omega\mu\beta\, \gamma^4 k^{-2}\, v = (i\omega\varepsilon)^{-1} J_b + q_3, \qquad (19\text{-}5a)$$
$$[\nabla^2 + \gamma^4 k^{-2}(1+k^2\beta^2)]\, v\ - 2i\omega\varepsilon\beta\, \gamma^4 k^{-2}\, u = (i\omega\mu)^{-1} K_b + q_4. \qquad (19\text{-}5b)$$

The various q-functions used in these equations are functions of the sources $\mathbf{J}$ and $\mathbf{K}$, and are given by

$$\nabla_t^2 q_1 = (i\omega\varepsilon)^{-1} \nabla_t \cdot \mathbf{J}_t, \tag{19-6a}$$

$$\nabla_t^2 q_2 = (i\omega\mu)^{-1} \nabla_t \cdot \mathbf{K}_t, \tag{19-6b}$$

$$\nabla_t^2 q_3 = -i\omega\mu\beta(\gamma/k)^2 \nabla_t \cdot (\mathbf{e}_b \times \mathbf{J}_t) - (i\omega\varepsilon)^{-1}(\partial/\partial x_b)\nabla_t \cdot \mathbf{J}_t$$
$$+ (\gamma/k)^2 \nabla_t \cdot (\mathbf{e}_b \times \mathbf{K}_t), \tag{19-6c}$$

$$\nabla_t^2 q_4 = -i\omega\varepsilon\beta(\gamma/k)^2 \nabla_t \cdot (\mathbf{e}_b \times \mathbf{K}_t) - (i\omega\mu)^{-1}(\partial/\partial x_b)\nabla_t \cdot \mathbf{K}_t$$
$$- (\gamma/k)^2 \nabla_t \cdot (\mathbf{e}_b \times \mathbf{J}_t). \tag{19-6d}$$

Equations (19-4) to (19-6) form the basis of the scalar treatment: if a solution of (19-5) is obtained, then the solution of (19-4) is also obtained.

In order to illustrate the procedure, consider a point electric dipole situated at the origin and radiating into the chiral medium; the electric source current density $\mathbf{J} = -i\omega\mathbf{e}_b\delta(\mathbf{r})$ and $\mathbf{J}_t = 0$. Then, after subscripting the Hertz potentials by the qualifier E, (19-5a,b) are of the forms

$$[\nabla^2 + \gamma^4 k^{-2}(1+k^2\beta^2)] u_E + 2i\omega\mu\beta \gamma^4 k^{-2} v_E = -\varepsilon^{-1} \delta(\mathbf{r}), \tag{19-7a}$$

$$[\nabla^2 + \gamma^4 k^{-2}(1+k^2\beta^2)] v_E - 2i\omega\varepsilon\beta \gamma^4 k^{-2} u_E = 0, \tag{19-7b}$$

and can be identically satisfied by

$$u_E = [\nabla^2 + \gamma^4 k^{-2}(1+k^2\beta^2)] w_E, \tag{19-8a}$$

$$v_E = 2i\omega\varepsilon\beta \gamma^4 k^{-2} w_E, \tag{19-8b}$$

provided the auxiliary function w_E is itself a solution of

$$[\nabla^2 + \gamma_1^2] [\nabla^2 + \gamma_2^2] w_E = -\varepsilon^{-1} \delta(\mathbf{r}). \tag{19-8c}$$

Utilizing the three-dimensional Fourier transforms, it can be easily shown that the solution of (19-8c) is given by

$$w_E = -(16\pi\varepsilon\beta\gamma^4 k^{-1})^{-1}[g(\gamma_1;\mathbf{r}) - g(\gamma_2;\mathbf{r})], \tag{19-9a}$$

where $g(\sigma;r)$ is defined by (13-13d), and whence,

$$u_E = (8\pi\varepsilon)^{-1}[g(\gamma_1;r) + g(\gamma_2;r)], \qquad (19\text{-}9b)$$

$$v_E = - (i\omega/8\pi k)[g(\gamma_1;r) - g(\gamma_2;r)], \qquad (19\text{-}9c)$$

by virtue of (19-8a,b), respectively. To find the fields radiated by an electric dipole $\mathbf{J} = -i\omega e_b\delta(\mathbf{r})$, (19-9) are substituted into (19-4); the result

$$\mathbf{E}(\mathbf{r}) = (\omega^2\mu/k)\,(\gamma/k)^2\,[\gamma_1\boldsymbol{\circledB}_1(\mathbf{r}, 0) + \gamma_2\boldsymbol{\circledB}_2(\mathbf{r}, 0)]\cdot e_b, \qquad (19\text{-}10a)$$

$$\mathbf{H}(\mathbf{r}) = -i\omega\,(\gamma/k)^2\,[\gamma_1\boldsymbol{\circledB}_1(\mathbf{r}, 0) - \gamma_2\boldsymbol{\circledB}_2(\mathbf{r}, 0)]\cdot e_b, \qquad (19\text{-}10b)$$

is identical to (15-5), demonstrating thereby the compatibility of Weiglhofer's scalar treatment with the vector methods.

As yet another illustration of the scalar treatment, consider a point magnetic dipole situated at the origin and radiating into the chiral medium; let the dipole orientation be e_b so that $\mathbf{K} = -i\omega e_b\delta(\mathbf{r})$ and $\mathbf{K_t} = \mathbf{0}$. Then, after subscripting the Hertz potentials by the qualifier M, equations (19-5a,b) are of the forms

$$[\nabla^2 + \gamma^4 k^{-2}(1+k^2\beta^2)]\,u_M\ + 2i\omega\mu\beta\,\gamma^4 k^{-2}\,v_M = 0, \qquad (19\text{-}11a)$$

$$[\nabla^2 + \gamma^4 k^{-2}(1+k^2\beta^2)]\,v_M\ - 2i\omega\varepsilon\beta\,\gamma^4 k^{-2}\,u_M = - \mu^{-1}\,\delta(\mathbf{r}), \qquad (19\text{-}11b)$$

which can be solved by the prescription of an auxiliary function w_M as

$$u_M = -2i\omega\mu\beta\,\gamma^4 k^{-2}w_M, \qquad (19\text{-}12a)$$

$$v_M = [\nabla^2 + \gamma^4 k^{-2}(1+k^2\beta^2)]\ w_M, \qquad (19\text{-}12b)$$

provided the auxiliary function w_M is itself a solution of

$$[\nabla^2 + \gamma_1^2]\,[\nabla^2 + \gamma_2^2]\,w_M = - \mu^{-1}\,\delta(\mathbf{r}). \qquad (19\text{-}12c)$$

The function w_M can be found by comparing (19-12c) with (19-8c), whence

$$u_M = (i\omega/8\pi k)[g(\gamma_1;r) - g(\gamma_2;r)], \qquad (19\text{-}13a)$$

$$v_M = (8\pi\mu)^{-1}[g(\gamma_1;r) + g(\gamma_2;r)]. \qquad (19\text{-}13b)$$

In order to find the fields radiated by the magnetic dipole $\mathbf{K} = -i\omega e_b\delta(\mathbf{r})$, (19-13) are substituted into (19-4); the result

$$\mathbf{E}(\mathbf{r}) = i\omega\,(\gamma/k)^2\,[\gamma_1\boldsymbol{\mathfrak{G}}_1(\mathbf{r},\,0) - \gamma_2\boldsymbol{\mathfrak{G}}_2(\mathbf{r},\,0)]\cdot e_b, \qquad (19\text{-}14a)$$

$$\mathbf{H}(\mathbf{r}) = (\omega^2\varepsilon/k)\,(\gamma/k)^2\,[\gamma_1\boldsymbol{\mathfrak{G}}_1(\mathbf{r},\,0) + \gamma_2\boldsymbol{\mathfrak{G}}_2(\mathbf{r},\,0)]\cdot e_b, \qquad (19\text{-}14b)$$

is the dual of (19-10) as per the duality relations (5-10).

20. ACOUSTICALLY CHIRAL SOLIDS

At the beginning of this work it was mentioned that chirality is due to the handedness in the geometry of the microstructure; this is not only characteristic of naturally optically active substances, but can also be utilized to build (artificially) chiral composites. Since the electromagnetic waves can discriminate between objects which are otherwise identical but have opposite handedness, there is no reason to suppose that other transverse waves cannot do so too. Using a helical ensemble of rigid beads suspended in a solid continuum, it has been shown [Varadan et al 1986] that acoustic waves in solids can also sense the handedness of a scatterer, if the scatterer is chiral. With the proliferation of research on multiphase composites, this suggests the viability of acoustically active composites. In what follows, the concept of chirality will be extended to acoustically chiral solids; the notation is slightly different from that of the preceding sections, but not by far.

First, a brief review of classical (achiral) elastodynamics. The Cauchy infinitesimal strain tensor $\mathfrak{E}$ in an isotropic, homogeneous solid is symmetric and is given in terms of the particle displacement vector $\mathbf{u}$ by

$$\mathfrak{E} = (1/2)[\nabla\mathbf{u} + \mathbf{u}\nabla]. \tag{20-1}$$

The stress tensor $\mathfrak{T}$ is also symmetric, and can be determined from $\mathfrak{E}$ via the relationship,

$$\mathfrak{T} = \lambda\mathfrak{J}[\text{trace}\,\mathfrak{E}] + 2\mu\mathfrak{E} = \lambda(\nabla\cdot\mathbf{u})\mathfrak{J} + \mu\,[\nabla\mathbf{u} + \mathbf{u}\nabla], \tag{20-2}$$

λ and μ being the Lamé constants. The constitutive equation (20-2) is in addition to the time-harmonic Newton's law of motion

$$\nabla\cdot\mathfrak{T} + \rho\omega^2\mathbf{u} + \rho\mathbf{F} = 0, \tag{20-3}$$

in which $\mathbf{F}$ is the body force per unit mass, and ρ is the mass density. Combining (20-2) and (20-3) leads to the usual Navier's equation for $\mathbf{u}$,

$$(\lambda + 2\mu)\nabla\nabla\cdot\mathbf{u} - \mu\nabla\times\nabla\times\mathbf{u} + \rho\omega^2\mathbf{u} + \rho\mathbf{F} = 0. \tag{20-4}$$

The solution of this equation gives rise to two fields: the first one is longitudinal and has a phase velocity equal to $[(\lambda + 2\mu)/\rho]^{1/2}$, and other one is transverse with the slower phase velocity $[\mu/\rho]^{1/2}$. There is no acoustic activity obtainable from classical elastodynamics [Lakhtakia et al 1988c].

In 1909 the Cosserat brothers published a theory of elasticity for media with oriented particles; this theory built upon the earlier assertion by Voigt [1887] that the interaction between two adjacent interior volume elements in the body was due both to a force vector and a moment vector. In the Cosserat theory, therefore, every material particle is associated with a perfectly rigid triad which, in the course of deformation, undergoes not only a displacement but also a rotation. Thus, composite media with fibrous microstructures have three additional degrees of freedom: the fibers can rotate as well, thereby setting the stage for the (additional) microrotation field $\mathbf{j}$. Consequently, the stress tensor is not symmetric. These developments have been eloquently described by Eringen [1968] in his treatment of what are now called micropolar elastic solids, and have also been comprehensively dealt with by W. Nowacki [1986]. Isotropic, micropolar solids have $\mathbf{u}$ and $\mathbf{j}$ coupled, thereby leading to dispersive wave velocities; nevertheless, they are not mirror-asymmetric.

The linear theory for noncentrosymmetric (hemitropic) micropolar solids was developed in the early 1960's [Kuvshinskii & Aero 1963; Aero & Kuvshinskii 1964], and the elastostatics of such media have also been examined. In particular, the propagation of longitudinal waves has been investigated [J.P. Nowacki 1977a] and the infinite-medium elastostatic Green's function has been obtained [J.P. Nowacki 1977b]. Lakes and Benedict [1982] have studied the torsion of hemitropic micropolar cylinders, and Lakhtakia et al [1988c] have derived approximate solutions for the radiation problem. It goes without saying that these developments can be applied to a fibrous composite with twisted or spiraling fibers.

Noncentrosymmetric, isotropic micropolar solids can be described by the constitutive equations [Lakes & Benedict 1982]:

$$\mathfrak{T} = \lambda(\nabla\cdot u)\mathbf{I} + (2\mu+\kappa)\mathfrak{E} - (\kappa/2)\mathbf{I}\times(\nabla\times u) + \kappa\mathbf{I}\times\mathbf{j}$$

$$+ C_1(\nabla \cdot \mathbf{j})\mathbf{I} + C_2\, \mathbf{j}\nabla + C_3\nabla\mathbf{j}, \tag{20-5}$$

$$\mathbf{M} = \alpha(\nabla \cdot \mathbf{j})\mathbf{I} + \beta\mathbf{j}\nabla + \gamma\nabla\mathbf{j} + C_1(\nabla \cdot \mathbf{u})\mathbf{I} + (C_2 + C_3)\mathbf{E}$$
$$- (C_3 - C_2)\mathbf{I}\times(\nabla\times\mathbf{u})/2 + (C_3 - C_2)\,\mathbf{I}\times\mathbf{j}, \tag{20-6}$$

where $\mathbf{M}$ is the couple stress tensor. There are some constraints [Lakes & Benedict 1982] on the material constants λ, μ, α, β, γ, κ, C_1, C_2 and C_3, which arise from the consideration that the internal energy be positive. By setting C_1, C_2 and C_3 to zero, the equations of micropolar elastic solids [Eringen 1968] are recovered, and this case has been extensively investigated by W. Nowacki [1986]; and, additionally, by setting α, β, γ and κ also equal to zero, the usual equations for ordinary elastic solids can be obtained. Not all of these constants are necessary for chirality, and it can be easily shown that C_1, C_2 as well as κ can be zero for noncentrosymmetric, isotropic solids. With this assumption, the corresponding equations of motion are given as [Lakhtakia et al 1988c]

$$(\lambda+2\mu)\nabla\nabla \cdot \mathbf{u} - \mu\nabla\times\nabla\times\mathbf{u} + \rho\omega^2\mathbf{u}$$
$$+ C_3\nabla\nabla \cdot \mathbf{j} - C_3\nabla\times\nabla\times\mathbf{j} + \rho\mathbf{F} = 0, \tag{20-7}$$
$$(\alpha+\beta+\gamma)\nabla\nabla \cdot \mathbf{j} - \gamma\nabla\times\nabla\times\mathbf{j} + \rho\omega^2\mathbf{j} + 2\,C_3\nabla\times\mathbf{j}$$
$$+ C_3\nabla\nabla \cdot \mathbf{u} - C_3\nabla\times\nabla\times\mathbf{u} + \rho\mathbf{L} = 0, \tag{20-8}$$

in which $\mathbf{L}$ is the body couple per unit mass. The constitutive properties used in the foregoing equations must satisfy the following inequalities:

$$3\lambda + 2\mu \geq 0, \qquad\qquad \mu \geq 0,$$
$$3\alpha + \beta + \gamma \geq 0, \qquad\qquad -\gamma \leq \beta \leq \gamma, \tag{20-9}$$
$$C_3{}^2 \leq 8\mu(\gamma + \beta), \qquad\qquad C_3{}^2 \leq 8\mu(\gamma - \beta),$$
$$C_3{}^2 \leq 4(3\lambda + 2\mu)(3\alpha + \beta + \gamma).$$

Let a planewave be propagating in the chiral elastic solid in the +z direction. Associated with the wave are the displacement and the microrotation vectors

$$\mathbf{u} = \mathbf{A}\, \exp[ikz]; \qquad \mathbf{j} = \mathbf{B}\, \exp[ikz], \tag{20-10}$$

with k being the wavenumber. Substitution of (20-10) in the homogeneous equations (20-7) and (20-8) leads to a separation of the longitudinal and the transverse fields. For the longitudinal components, it turns out that

$$[\rho\omega^2 - k^2(\lambda+2\mu)] A_z - [k^2 C_3] B_z = 0, \qquad (20\text{-}11a)$$
$$[\rho\omega^2 - k^2(\alpha+\beta+\gamma)] B_z - [k^2 C_3] A_z = 0, \qquad (20\text{-}11b)$$

which yields the dispersion relation

$$k^4[(\lambda+2\mu)(\alpha+\beta+\gamma) - C_3^2] - k^2\rho\omega^2[\lambda+2\mu+\alpha+\beta+\gamma] - \rho^2\omega^4 = 0. \quad (20\text{-}12)$$

Thus, two non-dispersive longitudinal planewaves can exist inside the noncentrosymmetric, isotropic medium; for each solution k^2 of (20-12), the ratio of amplitudes is given by

$$B_z/A_z = [\rho\omega^2 - k^2(\lambda+2\mu)]/k^2 C_3. \qquad (20\text{-}13)$$

The analysis for the transverse components is more complicated. On eliminating A_x and A_y, the dispersion equation obtained is given by

$$[k^4(C_3^2 - \gamma\mu) + k^2\rho\omega^2(\gamma+\mu) - \rho^2\omega^4]^2 - [2kC_3(\mu k^2 - \rho\omega^2)]^2 = 0, \quad (20\text{-}14)$$

which equation is quartic in k^2. Thus, four separate transverse fields are possible, whose phase velocities are generally dispersive; for each solution k^2 of (20-14), the ratios of amplitudes are given by

$$B_x/A_x = [\rho\omega^2 - \mu k^2]/k^2 C_3, \qquad (20\text{-}15a)$$
$$A_x/A_y = B_x/B_y = i\,[2kC_3(\mu k^2 - \rho\omega^2)]$$
$$\bullet[k^4(C_3^2 - \gamma\mu) + k^2\rho\omega^2(\gamma+\mu) - \rho^2\omega^4]^{-1}. \quad (20\text{-}15b)$$

As can be seen from (20-14), the ratio on the right hand side of (20-15b) must equal $\pm i$; hence two of the transverse waves must be LCP, while the remaining two are RCP.

Equations (20-7) and (20-8) can be manipulated into more meaningful forms if a separation of the longitudinal and transverse components is carried. Let there be a set of sources $\{F_p, L_p\}$ such that only the longitudinal fields $\{u_p, j_p\}$ exist; $\nabla \times u_p = 0$ and $\nabla \times j_p = 0$. In that case these two equations reduce to

$$(\lambda + 2\mu) \nabla\nabla \cdot u_p + \rho\omega^2 u_p + C_3 \nabla\nabla \cdot j_p + \rho F_p = 0, \qquad (20\text{-}16a)$$

$$(\alpha + \beta + \gamma)\nabla\nabla \cdot j_p + \rho\omega^2 j_p + C_3 \nabla\nabla \cdot u_p + \rho L_p = 0. \qquad (20\text{-}16b)$$

Elimination of u_p from (20-16a,b) leads to the differential equation

$$\mathcal{P} \cdot j_p = -\rho[(\lambda + 2\mu)\nabla\nabla \cdot L_p + \rho\omega^2 L_p - C_3\nabla\nabla \cdot F_p]; \qquad (20\text{-}17a)$$

while that of j_p yields

$$\mathcal{P} \cdot u_p = -\rho[(\alpha + \beta + \gamma)\nabla\nabla \cdot F_p + \rho\omega^2 F_p - C_3\nabla\nabla \cdot L_p]. \qquad (20\text{-}17b)$$

In these two equations, the differential operator $\mathcal{P}$ is defined by

$$\mathcal{P} = [(\alpha + \beta + \gamma)(\lambda + 2\mu) - C_3^2]\nabla\nabla \cdot \nabla\nabla$$
$$+ \rho\omega^2[\alpha + \beta + \gamma + \lambda + 2\mu]\nabla\nabla + \rho^2\omega^4 \mathbf{J}; \qquad (20\text{-}18)$$

and as a result, the solutions of (20-17a,b) can be obtained quite simply as

$$j_p = -\rho\, \mathcal{P}^{-1} \cdot [(\lambda + 2\mu) \nabla\nabla \cdot L_p + \rho\omega^2 L_p - C_3\nabla\nabla \cdot F_p], \qquad (20\text{-}19a)$$

$$u_p = -\rho\, \mathcal{P}^{-1} \cdot [(\alpha + \beta + \gamma)\nabla\nabla \cdot F_p + \rho\omega^2 F_p - C_3\nabla\nabla \cdot L_p]. \qquad (20\text{-}19b)$$

In a similar fashion, let a set of sources $\{F_s, L_s\}$ give rise to the fields $\{u_s, j_s\}$ which are purely solenoidal, i.e., $\nabla \cdot u_s = 0$ and $\nabla \cdot j_s = 0$. In that case, equations (20-7) and (20-8), respectively, reduce to

$$-\mu\nabla \times \nabla \times u_s + \rho\omega^2 u_s - C_3\nabla \times \nabla \times j_s + \rho F_s = 0, \qquad (20\text{-}20a)$$

$$-\gamma\nabla \times \nabla \times j_s + \rho\omega^2 j_s + 2C_3\nabla \times j_s - C_3\nabla \times \nabla \times u_s + \rho L_s = 0. \qquad (20\text{-}20b)$$

Elimination of $\mathbf{u_s}$ from the foregoing expressions can then be carried out, or that of $\mathbf{j_s}$: the field equations thus obtained can be compactly set down as

$$\mathbf{j_s} = \rho\mathcal{S}^{-1}\cdot [-\mu\nabla\times\nabla\times\mathbf{L_s} + \rho\omega^2\mathbf{L_s} + C_3\nabla\times\nabla\times\mathbf{F_s}], \tag{20-21a}$$

$$\mathbf{u_s} = \rho\mathcal{S}^{-1}\cdot [-\gamma\nabla\times\nabla\times\mathbf{F_s} + \rho\omega^2\mathbf{F_s} + 2C_3\nabla\times\mathbf{F_s} + C_3\nabla\times\nabla\times\mathbf{L_s}], \tag{20-21b}$$

where the dyadic differential operator $\mathcal{S}$ is given as

$$\mathcal{S} = [C_3{}^2 - \gamma\mu]\nabla\times\nabla\times\nabla\times\nabla\times\nabla\times\mathbf{I} + 2\mu C_3\nabla\times\nabla\times\nabla\times\mathbf{I}$$
$$+ \rho\omega^2(\gamma+\mu)\nabla\times\nabla\times\mathbf{I} - 2\rho\omega^2 C_3\nabla\times\mathbf{I} - \rho^2\omega^4\mathbf{I}. \tag{20-22}$$

The derivation procedure of the inverses $\mathcal{P}^{-1}$ and $\mathcal{S}^{-1}$ is very straightforward, and can be easily implemented using the techniques described in Sections 15 and 16. It is the solution of the dispersion equation (20-14) which is not tractable, except computationally, due to its significant complexity. Hence, the approximation $C_3{}^2 << \gamma\mu$ is suggested, which may not be entirely invalid in view of the conditions (20-9) as well as previous experience in electromagnetics. With this approximation, the definitions (20-18) and (20-22) simplify to

$$\mathcal{P} \cong [(\alpha+\beta+\gamma)(\lambda+2\mu)]\nabla\nabla\cdot\nabla\nabla + \rho\omega^2[\alpha+\beta+\gamma+\lambda+2\mu]\nabla\nabla$$
$$+ \rho^2\omega^4\mathbf{I}; \tag{20-23}$$

$$\mathcal{S} \cong -\gamma\mu\nabla\times\nabla\times\nabla\times\nabla\times\mathbf{I} + 2\mu C_3\nabla\times\nabla\times\nabla\times\mathbf{I}$$
$$+ \rho\omega^2(\gamma+\mu)\nabla\times\nabla\times\mathbf{I} - 2\rho\omega^2 C_3\nabla\times\mathbf{I} - \rho^2\omega^4\mathbf{I}, \tag{20-24}$$

and it should be noted that the parameter C_3 has completely disappeared in (20-23), but not from (20-24). Then, the dyadic inverses can be worked out to be

$$4\pi\rho^2\omega^4\mathcal{P}^{-1} \cong \mathbf{I}\delta(\mathbf{r}) + [1 - (\lambda+2\mu)/(\alpha+\beta+\gamma)]^{-1} \nabla\nabla g(k_1;r)$$
$$+ [1 - (\alpha+\beta+\gamma)(\lambda+2\mu)]^{-1} \nabla\nabla g(k_2;r), \tag{20-25}$$

$$-4\pi\mathcal{S}^{-1} \cong [\mu\gamma^2(k_3{}^2 - k_4{}^2)(k_3{}^2 - k_5{}^2)]^{-1}$$
$$\cdot \{(\gamma k_3{}^2-\rho\omega^2)(\mathbf{I} + \nabla\nabla/k_3{}^2) + 2C_3\nabla\times\mathbf{I}\} g(k_3;r)$$

$$+ [2\mu(k_4^2 - k_3^2)\sqrt{(C_3^2 + \gamma\rho\omega^2)}]^{-1}\{k_4\mathbf{j} + \nabla\nabla/k_4 - \nabla\times\mathbf{j}\}g(k_4;r)$$

$$+ [2\mu(k_5^2 - k_3^2)\sqrt{(C_3^2 + \gamma\rho\omega^2)}]^{-1}\{k_5\mathbf{j} + \nabla\nabla/k_5 + \nabla\times\mathbf{j}\}g(k_5;r). \quad (20\text{-}26)$$

with $g(\sigma;r) = \exp[i\sigma r]/r$ as per (13-13d). In these two equations (20-25) and (20-26), the wavenumbers

$$k_1^2 = \rho\omega^2/(\alpha+\beta+\gamma), \qquad\qquad\qquad\qquad (20\text{-}27a)$$
$$k_2^2 = \rho\omega^2/(\lambda+2\mu), \qquad\qquad\qquad\qquad (20\text{-}27b)$$
$$k_3^2 = \rho\omega^2/\mu, \qquad\qquad\qquad\qquad\qquad (20\text{-}27c)$$

are non-dispersive, whereas the remaining two are dispersive and can be obtained from

$$\gamma k_4 = \sqrt{(C_3^2 + \gamma\rho\omega^2)} - C_3, \qquad\qquad\qquad (20\text{-}27d)$$
$$\gamma k_5 = \sqrt{(C_3^2 + \gamma\rho\omega^2)} + C_3. \qquad\qquad\qquad (20\text{-}27e)$$

Given the approximation $[\gamma\mu - C_3^2] \cong \gamma\mu$, from (20-21), (20-25) and (20-26) it can be concluded that there are five eigenmodes. Each of the five modes has a displacement vector $\mathbf{u}$ and a microrotation vector $\mathbf{j}$ associated with it. The first two (k_1 and k_2) are longitudinal with non-dispersive phase velocities. The last two (k_4 and k_5) are transverse with dispersive phase velocities; respectively, they are right- and left- circularly polarized. Since right- (resp. left-) handed media have $C_3 > 0$ (resp. < 0), the RCP (resp. LCP) eigenwave possesses the higher phase velocity of the two. The status of the third wave (k_3) is not quite so transparent due to the presence of C_3 in the prefactor of $g(k_3;r)$ in (20-26), except that it is clearly transverse.

In order to examine the effect of the approximation $[\gamma\mu - C_3^2] \cong \gamma\mu$, we consider planewave propagation once again. Then from (20-11) it is clear that k_1 is the wavenumber of a longitudinal $\mathbf{j}$-wave, and k_2 is that of a longitudinal $\mathbf{u}$-wave. For the transverse wave with wavenumber k_3, the right side of (20-15a) simplifes to $0/k_3^2C_3$, while the right side of (20-15b) assumes the indeterminate form $0/0$; as a consequence this is a linearly-polarized u-wave with no microrotation field attached to it. The fourth wave (k_4) is a transverse RCP wave with both

106

displacement and microrotation fields attached to it; this is because the right side of (20-15a) is non-zero while the polarization ratios $A_x/A_y = B_x/B_y = i$ from (20-15b). Finally, the fifth wave (k_5) is a transverse LCP wave with both displacement and microrotation fields associated with it, since the right side of (20-15a) is non-zero while the polarization ratios $A_x/A_y = B_x/B_y = -i$ from (20-15b). In view of the fact that both k_4 and k_5 are related to γ and not to μ, the last two circularly-polarized planewaves should perhaps be labelled as quasi-microrotation waves.

21. SELECTED DYADIC RELATIONS

§§ Let $\mathfrak{A}$ be an arbitrary dyadic, $\mathfrak{I}$ is the idempotent, the lower case bold letters represent vectors, and φ is a scalar function. Some pertinent relations from dyadic analysis are given below:

§ $u\times\mathfrak{I} = \mathfrak{I}\times u.$

§ $\nabla\bullet(uv) = (\nabla\bullet u)v + (u\bullet\nabla)v.$

§ $\nabla\times(u\times v) = \nabla\bullet(vu - uv).$

§ $\nabla\bullet(u\times\mathfrak{A}) = (\nabla\times u)\bullet A - u\bullet(\nabla\times\mathfrak{A}).$

§ $\nabla\bullet\nabla\times\mathfrak{A} = 0.$

§ $\nabla(\varphi u) = (\nabla\varphi)u + \varphi\nabla u.$

§ $\nabla(u\bullet v) = (\nabla u)\bullet v + (\nabla v)\bullet u.$

§ $\nabla\times(uv) = (\nabla\times u)v - u\times\nabla v.$

§ $\nabla\times(\nabla u) = 0.$

§ $\nabla(u\times v) = (\nabla u)\times v - (\nabla v)\times u.$

§ $\nabla\times(\varphi\mathfrak{A}) = \nabla\varphi\times\mathfrak{A} + \varphi\nabla\times\mathfrak{A}.$

§ $\nabla\bullet[u\times(\nabla\times\mathfrak{A}) + (\nabla\times u)\times\mathfrak{A}] = (\nabla\times\nabla\times u)\bullet\mathfrak{A} - u\bullet(\nabla\times\nabla\times\mathfrak{A}).$

§ $\nabla\bullet(\varphi\mathfrak{I}) = \nabla\varphi.$

§ $\nabla\times(\varphi\mathfrak{I}) = \nabla\varphi\times\mathfrak{I}.$

§ $\nabla\times\nabla\times(\varphi\mathfrak{I}) = \nabla\nabla\varphi - (\nabla^2\varphi)\mathfrak{I}.$

§ $\nabla\bullet(\mathfrak{I}\times u) = \nabla\times u.$

§§ The dyadic inverse of $\mathfrak{A}$ is given by $\mathfrak{A}^{-1} = [\text{adj } \mathfrak{A}]\,[|\mathfrak{A}|]^{-1}$, so that $\mathfrak{A}\bullet\mathfrak{A}^{-1} = \mathfrak{I} = \mathfrak{A}^{-1}\bullet\mathfrak{A}$; adj $\mathfrak{A}$ is the adjoint, and $|\mathfrak{A}|$ is the determinantal value, of $\mathfrak{A}$. Some useful cases follow:

§ If the dyadic $\mathfrak{A} = \lambda\mathfrak{I} + uv$, then

  $|\mathfrak{A}| = \lambda^2(\lambda + u\bullet v)$

  adj $\mathfrak{A} = \lambda^{-1}|\mathfrak{A}|\,\mathfrak{I} - \lambda uv.$

§ If the dyadic $\mathfrak{A} = \lambda\mathfrak{I} + uv\bullet\mathfrak{B}$, then

  $|\mathfrak{A}| = \lambda^2(\lambda + v\bullet\mathfrak{B}\bullet u)$

  adj $\mathfrak{A} = \lambda^{-1}|\mathfrak{A}|\,\mathfrak{I} - \lambda uv\bullet\mathfrak{B}.$

§ If the dyadic $A = \lambda \mathfrak{I} + \mathbf{c} \times \mathfrak{I}$, then

$$|\mathfrak{A}| = \lambda(\lambda^2 + \mathbf{c} \cdot \mathbf{c}),$$

$$\text{adj } \mathfrak{A} = \lambda(\lambda \mathfrak{I} - \mathbf{c} \times \mathfrak{I}) + \mathbf{cc}.$$

§If the dyadic $\mathfrak{A} = \lambda \mathfrak{I} + \mathbf{ab} + \mathbf{c} \times \mathfrak{I}$, then

$$|\mathfrak{A}| = \lambda^3 + \lambda^2(\mathbf{a} \cdot \mathbf{b}) + \lambda(\mathbf{c} \cdot \mathbf{c} - \mathbf{a} \times \mathbf{b} \cdot \mathbf{c}) + (\mathbf{a} \cdot \mathbf{c})(\mathbf{b} \cdot \mathbf{c})$$

$$\text{adj } \mathfrak{A} = \lambda^2 \mathfrak{I} - \lambda \mathbf{ab} + \lambda(\mathbf{a} \cdot \mathbf{b})\mathfrak{I} - \lambda \mathbf{c} \times \mathfrak{I} + \mathbf{cc} - (\mathbf{b} \cdot \mathbf{c})(\mathbf{a} \times \mathfrak{I}) - \mathbf{c}(\mathbf{a} \times \mathbf{b}).$$

22. SELECTED BIBLIOGRAPHY

Aero, E.L. and E.V. Kuvshinskii [1964] Continuum theory of asymmetric elasticity. Equilibrium of isotropic body, *Fizika Twërdogo Tela* **6**, 2689.

Agranovich, V.M. and V.L. Ginzburg [1973] Phenomenological electrodynamics of gyrotropic media, *Sov. Phys. JETP* **36**, 440.

Altman, C., A. Schatzberg and K. Suchy [1984] Symmetry transformations and reversal of currents and fields in bounded (bi)anisotropic media, *IEEE Antennas Propagat.* **32**, 1204.

Applequist, J. [1987] Optical activity: Biot's bequest, *American Scientist* **75**(1), 59.

Arago, F. [1811] Mémoire sur une modification remarquable qu'éprouvent les rayons lumineux dans leur passage à travers certains corps diaphanes, et sur quelques autres nouveaux phénomènes d'optique, in *Mémoires de la classe des sciences mathématiques et physiques de l'Institut impérial de France, Part 1*, 93.

Astrov, D.N. [1960] The magnetoelectric effect in antiferromagnetics, *Zh. Eksp. Teor. Fiz.* **38**, 984.

Barber, P.W. and C. Yeh [1975] Scattering of electromagnetic waves by arbitrarily shaped dielectric bodies, *Appl. Opt.* **14**, 2864.

Barron, L.D. [1982] *Molecular Light Scattering and Optical Activity*, Cambridge, U.K.: C.U.P.

Barron, L.D., M.P. Bogaard and A.D. Buckingham [1973] Raman scattering of circularly polarized light by optically active molecules, *J. Am. Chem. Soc.* **95**, 604.

Bassiri, S., N. Engheta and C.H. Papas [1986] Dyadic Green's function and dipole radiation in chiral media, *Alta Freq.* **55**, 83.

Bassiri, S., C.H. Papas and N. Engheta [1988] Electromagnetic wave propagation through a dielectric-chiral interface and through a chiral slab, *J. Opt. Soc. Am. A* **5**, 1450.

Belmont, A.S., S. Zietz and C. Nicolini [1985] Differential scattering of circularly polarized light by chromatin modeled as a helical array of dielectric ellipsoids within the Born approximation, *Biopolymers* **24**, 1301.

Biot, J.B. [1812] Mémoire sur un noveau genre d'oscillation que les molécules de la lumière éprouvent en traversant certains cristaux, in *Mémoires de la classe des sciences mathématiques et physiques de l'Institut impérial de France, Part 1*, 1.

Biot, J.B. [1817] Mémoire sur les rotations que certains substances impriment aux axes de polarisation des rayons lumineux, *Mémoires de l'Académie royale des sciences de l'Institut de France* **2**, 41.

Biot, J.B. [1835] Mémoire sur la polarisation circulaire et sur les applications à la chimie organique, *Mémoires de l'Académie royale des sciences de l'Institut de France* **13**, 39.

Birss, R.R. and R.G. Shrubsall [1967] The propagation of EM waves in magnetoelectric crystals, *Phil. Mag.* **15**, 687.

Bohren, C.F. [1974] Light scattering by an optically active sphere, *Chem. Phys. Lett.* **29**, 458.

Bohren, C.F. [1975a] *Light scattering by optically active particles*, Ph.D. dissertation (U. Arizona, Tucson).

Bohren, C.F. [1975b] Scattering of electromagnetic waves by an optically active spherical shell, *J. Chem. Phys.* **62**, 1566 (1975).

Bohren, C.F. [1976] Angular dependence of the scattering contribution to circular dichroism, *Chem. Phys. Lett.* **40**, 391.

Bohren, C.F. [1978] Scattering of electromagnetic waves by an optically active cylinder, *J. Colloid Interface Sci.* **66**, 105 (1978).

Bohren, C.F. and D.R. Huffman [1983] *Absorption and Scattering of Light by Small Particles*, New York: Wiley.

Bokut', B.V. and F.I. Fedorov [1959] On the theory of optical activity in crystals. III. General equations of normals, *Opt. Spectrosc.* **6**, 342.

Bokut', B.V. and F.I. Fedorov [1960] Reflection and refraction of light in optically isotropic active media, *Opt. Spectrosc.* **9**, 334.

Bokut', B.V. and B.A. Sotskii [1963] The passage of light through an optically active absorbing plate, *Opt. Spectrosc.* **14**, 60.

Bokut', B.V. and Serdyukov [1972] On the phenomenological theory of natural optical activity, *Sov. Phys. JETP* **34**, 962.

Born, M. [1915] Über die natürliche optische Aktivität von Flüssigkeiten und Gasen, *Phys. Z* **16**, 251.

Born, M. [1972] *Optik*, Heidelberg: Springer Verlag.

Bouchiat, M. and L. Pottier [1986] Optical experiments and weak interactions, *Science* **234**, 1203.

Bour, L.J. and N.J. Lopes Cardozo [1981] On the birefringence of the living human eye, *Vision Res.* **21**, 1413.

Brand, D.J. and J. Fisher [1987] Molecular structure and chirality, *J. Chem. Educ.* **64**, 1035 (1987).

Buchwald, J.Z. [1985] *From Maxwell to Microphysics*, Chicago: U. Chicago Press (1985).

Bunkin, F.V. [1957] On radiation in anisotropic media, *Sov. Phys. JETP* **5**, 277.

Bustamente C., I. Tinoco, Jr. and M.F. Maestre [1982] Cicrcular intensity differential scattering of light. IV. Randomly oriented species, *J. Chem. Phys.* **76**, 3440.

Chambers, Ll.G. [1956] Propagation in a gyrational medium, *Q. J. Mech. Appl. Math.* **9**, 360.

Charney, E. [1979] *The Molecular Basis of Optical Activity*, Malabar, FL: Krieger.

Chawla, B.R. and H. Unz [1971] *Electromagnetic Waves in Moving Magneto-Plasmas*, Lawrence: U.P. Kansas.

Chen, H.C. [1983] *Theory of Electromagnetic Waves*, New York: McGraw-Hill.

Cheng, D.K. and J.A. Kong [1968a] Covariant descriptions of bianisotropic media, *Proc. IEEE* **56**, 248.

Cheng, D.K. and J.A. Kong [1968b] Time-harmonic fields in source-free bianisotropic media, *J. Appl. Phys.* **39**, 5792.

Chow, Y. [1962] A note on radiation in a gyro-electric-magnetic medium -- An extension of Bunkin's calculation, *IRE Trans. Antennas Propagat.* **10**, 464.

Cochran, W., F.H.C. Crick and V. Vand [1952] The structure of synthetic polypeptides. I. The transform of atoms on a helix, *Acta. Cryst.* **5**, 581.

Condon, E.U. [1937] Theories of optical rotatory power, *Rev. Mod. Phys.* **9**, 432.

Corley, L.S. and O. Vogl [1980] Optically active polychloral, *Polymer Bull.* **3**, 211.

Cosserat, E. and F. Cosserat [1909] *Théorie des Corps Déformables*, Paris: A. Hermann et Fils.

Crowe, K., J. Duclos, G. Fiorentini and G. Torelli (Eds.) [1980] *Fundamental Interactions and the Structure of Matter Vol. 57*, New York: Plenum.

de Hoop, A.T. [1960] A reciprocity theorem for the electromagnetic field scattered by an obstacle, *Appl. Sci. Res. B* **8**, 135.

de Hoop, A.T. [1959] On the plane-wave extinction cross-section of an obstacle, *Appl. Sci. Res. B* **7**, 463.

deVries, H.L., A. Spoor and R. Jielof [1953] Properties of the eye with respect to polarised light, *Physica* **19**, 419.

Drude, P. [1900] *Lehrbuch der Optik*, Leipzig: S. Hirzel.

Elert, M.L. and C.T. White [1987] *Macromolecules* **20**, 1411.

Engheta, N. and A.R. Mickelson [1982] Transition radiation caused by a chiral plate, *IEEE Trans. Antennas Propagat.* **30**, 1213.

Eringen, A.C. [1968] Theory of micropolar elastic solids, in *Fracture II* [Ed.: H. Liebowitz], New York: Academic.

Eringen, A.C. [1984] Electrodynamics of memory-dependent nonlocal elastic continua, *J. Math. Phys.* **25**, 3235.

Eyring, H., J. Walter and G.E. Kimball [1944] *Quantum Chemistry*, New York: Wiley.

Fedorov, F.I. [1959a] On the theory of optical activity in crystals. I. The law of conservation of energy and the optical activity tensors, *Opt. Spectrosc.* **6**, 49.

Fedorov, F.I. [1959b] On the theory of optical activity in crystals. II. Crystals of cubic symmetry and plane classes of central symmetry, *Opt. Spectrosc.* **6**, 237.

Fresnel, A.J. [1866] *Oeuvres complétes*, Paris: Imprimerie impériale (1866).

Fuchs, R. [1965] Wave propagation in a magnetoelectric medium, *Phil. Mag.* **11**, 647.

Gordon, D.J. [1972] Mie scattering by optically active particles, *Biochem.* **11**, 413.

Gordon, D.J. and G. Holzwarth [1971] Optical activity of membrane suspensions: Calculation of artifacts by Mie scattering theory, *Proc. Nat. Acad. Sci. USA* **68**, 2365.

Gray, F. [1916] The optical activity of liquids and gases, *Phys. Rev.* **7**, 472.

Guire, T., M. Umari, V.V. Varadan and V.K. Varadan [1988] Microwave measurements on chiral composites, *Abst. URSI Radio Sci. Meet.* (Syracuse Univ., June 6-10).

Hansen, A.E. and T.D. Bouman [1980] Natural chiroptical spectroscopy: Theory and computations, *Adv. Phys. Chem.* **44**, 545.

Haracz, R.D., L.D. Cohen, A. Cohen and R.T. Wang [1987] Scattering of linearly polarized microwave radiation from a dielectric helix, *Appl. Opt.* **26**, 4632.

Harrington, R.F. [1964] *Time-Harmonic Electromagnetic Fields*, New York: McGraw-Hill.

Harrington, R.F. [1968] *Field Computation by Moment Methods*, New York: McGraw-Hill.

Harris, W.J. and O. Vogl [1981] Synthesis of optically active polymers, *Polymer Preprints* **22**, 309.

Hegstrom, R.A., J.P. Chamberlain, K. Seto and R.G. Watson [1988] Mapping the weak chirality of atoms, *Am. J. Phys.* **56**, 1086.

Heppke, G., D. Lötzsch and F. Oestreicher [1986] Chirale dotierstoffe mit außerwöhnlich hohem verdrillungsvermögen, *Z. Naturforsch.* **41A**, 1214.

Holzwarth, G., D.G. Gordon, J.E. McGiness, B.P. Dorman and M.F. Maestre [1974] Mie scattering contributions to the optical density and circular dichroism of T2 bacteriophage, *Biochem.* **13**, 126.

Hornreich, R.M. and S. Shtrikman [1968] Theory of gyrotropic birefringence, Phys. Rev. **171**, 1065.

Hug, W., S. Kint, G. Bailey and J. Scherer [1975] Raman circular intensity differential spectroscopy, *J. Am. Chem. Soc.* **97**, 5589.

Illert, C. [1987] Formulation and solution of the classical seashell problem. I. Seashell geometry, *Nuovo Cimento D* **9**, 791.

Ishimaru, A. [1978] *Wave Propagation and Scattering in Random Media, Vols. I & II*, New York: Academic.

Islam, M.N., A. Kponou, B. Suleman and W. Happer [1981] Magnetic circular dichroism of excimer molecules, *Phys. Rev. Lett.* **47**, 643.

Jackson, J.D. [1975] *Classical Electrodynamics*, New York: Wiley.

Jaggard, D.L., A.R. Mickelson and C.H. Papas [1978] On electromagnetic waves in chiral media, *Appl. Phys.* **18**, 211.

Jaggard, D.L., X. Sun and N. Engheta [1988] Canonical sources and duality in chiral media, *IEEE Trans. Antennas Propagat.* **36**, 1007.

Jerrard, H.G. [1954] Transmission of light through birefringent and optically active media: the Poincaré sphere, *J. Opt. Soc. Am.* **44**, 86.

Johnk, C.T.A. [1975] *Enginering Electromagnetic Fields and Waves*, New York: Wiley.

Kastener, R. and R. Mittra [1983] A spectral-iteration technique for analyzing scattering from arbitrary bodies. I, *IEEE Trans. Antennas Propagat.* **31**, 499.

Keller, D., C. Bustamente, M.F. Maestre and I. Tinoco, Jr. [1985] Model computations on the differential scattering of circularly polarized light by dense macromolecular particles, *Biopolymers* **24**, 783.

Kerker, M. [1969] *The Scattering of Light and Other Electromagnetic Radiation*, New York: Academic.

Kibble, T.W.B. [1965] Conservation laws for free fields, *J. Math. Phys.* **6**, 1022.

Kogelnik, H. [1960] On electromagnetic radiation in magnetoionic media, *J. Res. NBS* **64D**, 515.

Kong, J.A. [1970] Reciprocity relationships for bianisotropic media, *Proc. IEEE* **58**, 1966.

Kong, J.A. [1971] Charged particles in bianisotropic media, *Radio Sci.* **6**, 1015.

Kong, J.A. [1972] Theorems of bianisotropic media, *Proc. IEEE* **60**, 1036.

Kong, J.A. [1974] Optics of bianisotropic media, *J. Opt. Soc. Am.* **64**, 1304.

Kong, J.A. and D.K. Cheng [1968a] Wave reflections from a conducting surface with a moving uniaxial sheath, *IEEE Trans. Antennas Propagat.* **16**, 577.

Kong, J.A. and D.K. Cheng [1968b] Wave behavior at an interface of a semi-infinite moving anisotropic medium, *J. Appl. Phys.* **39**, 2282.

Krowne, C.M. [1984] Electromagnetic theorems for complex anisotropic medium, *IEEE Trans. Antennas Propagat.* **32**, 1224.

Kuvshinskii, E.V. and E.L. Aero [1963] Continuum theory of asymmetric elasticity, *Fizika Twërdogo Tela* **5**, 2592.

Lakes, R.S. [1987] Foam structures with a negative Poisson's ratio, *Science* **235,** 1038.

Lakes, R.S. and R.L. Benedict [1982] Noncentrosymmetry in micropolar elasticity, *Int. J. Engng. Sci.* **20,** 1161.

Lakhtakia, A., V.K. Varadan and V.V. Varadan [1985] Scattering and absorption characteristics of lossy dielectric, chiral, nonspherical objects, *Appl. Opt.* **24,** 4146.

Lakhtakia, A., V.V. Varadan and V.K. Varadan [1986] A parametric study of microwave reflection characteristics of a planar achiral-chiral interface, *IEEE Trans. Electromag. Compat.* **28,** 90.

Lakhtakia, A., V.V. Varadan and V.K. Varadan [1987a] On the influence of chirality on the scattering response of a chiral scatterer, *IEEE Trans. Electromag. Compat.* **29,** 70.

Lakhtakia, A., V.K. Varadan and V.V. Varadan [1987b] Regarding the sources of radiation fields in an isotropic chiral medium, *J. Wave-Mater. Interact.* **2,** 183.

Lakhtakia, A., V.V. Varadan and V.K. Varadan [1988a] Field equations, Huygens's principle, integral equations, and theorems for radiation and scattering of electromagnetic waves in isotropic chiral media, *J. Opt. Soc. Amer. A* **5,** 175.

Lakhtakia, A., V.V. Varadan and V.K. Varadan [1988b] Radiation by a straight thin-wire antenna embedded in an isotropic chiral medium, *IEEE Trans. Electromag. Compat.* **30,** 84.

Lakhtakia, A., V.V. Varadan and V.K. Varadan [1988c] Elastic wave propagation in noncentrosymmetric, isotropic media: Dispersion and field equations, *J. Appl. Phys.* **63,** 5246.

Lakhtakia, A., V.K. Varadan and V.V. Varadan [1988d] Excitation of a planar achiral/chiral interface by near fields, *J. Wave-Mater. Interact.* **3,** 231.

Le Bel, J.A. [1874] Sur les relations qui existent entre les formules atomiques des corps organiques et le pouvoir rotatoire de leures dissolutions, *Bull. Soc. Chimique Paris* **22,** 337.

Lipkin, D.M. [1964] Existence of a new conservation law in electromagnetic theory, *J. Math. Phys.* **5,** 696.

Lindman, K. [1920] Über die durch ein isotropes System von Spiralförmigen Resonatoren erzeugte Rotationspolarisation der elektromagnetischen Wellen, *Ann. Physik* **63**, 621.

Lindman, K. [1922] Über die durch ein aktives Raumgitter erzeugte Rotations-polarisation der elektromagnetischen Wellen, *Ann. Physik* **69**, 270.

Mayes, P.E. [1958] The equivalence of electric and magnetic sources, *IRE Trans. Antennas Propagat.* **6**, 295.

Miyazawa, T. [1961] Molecular vibrations and structures of high polymers. II. *J. Polym. Sci.* **55**, 215.

Morse, P.M. and H. Feshbach [1953] *Methods of Theoretical Physics*, New York: McGraw-Hill.

Nowacki, J.P. [1977a] Some dynamical problems of hemitropic micropolar continuum, *Bull. Acad. Polo. Sci., Ser. Sci. Tech.* **25**, 465.

Nowacki, J.P. [1977b] Green functions for a hemitropic micropolar continuum, *Bull. Acad. Polo. Sci., Ser. Sci. Tech.* **25**, 619.

Nowacki, J.P. and W. Nowacki [1977] Some problems of hemitropic micropolar continuum, *Bull. Acad. Polo. Sci., Ser. Sci. Tech.* **25**, 297.

Nowacki, W. [1986] *Theory of Asymmetric Elasticity*, Oxford: Pergamon.

O'Dell, T.H. [1970] *The Electrodynamics of Magneto-electric Media*, Amsterdam: North-Holland.

Odijk, T. [1987] Pitch of a polymer cholestric, *J. Phys. Chem.* **91**, 6060.

O'Raifeartaigh, L. [1975] Spontaneous symmetry breaking for chiral scalar superfields, *Nucl. Phys. B* **96**, 331.

Oseen, C.W. [1915] Über die Wechselwirkung zwischen zwei elektrischen Dipolen und über die Drehung der Polarisationsebene in Kristallen und Flüssigkeiten, *Ann. Phys.* **48**, 1.

Pasteur, L. [1848] Sur les relations qui peuvent exister entre la forme cristalline, la composition chimique et le sens de la polarisation rotatoire, *Annales de chimie et de physique* **24**, 442.

Pasteur, L. [1850] Recherces sur les propriétés spécifiques de deux acides qui composent l'acide racémique, *Annales de chimie et de physique* **28**, 56.

Pattanayak, D.N. and J.L. Birman [1981a] Wave propagation in optically active and magnetoelectric media of arbitrary geometry, *Phys. Rev. B* **24**, 4271.

Pattanayak, D.N., A. Puri and J.L. Birman [1981b] Phenomenological electro-dynamics of bounded gyrotropic media near a dipole transition frequency, *Phys. Rev. B* **24**, 4291.

Patterson, C.W., S.B. Singham and G.C. Salzman [1986] Circular intensity differential scattering of light by hierarchial molecular structures, *J. Chem. Phys.* **84**, 1916.

Piez, K.A. [1984] Molecular and aggregate structures of the collagens, in *Extra-cellular Matrix Biochemistry* [Eds.: K.A. Piez and A.H. Reddi], New York: Elsevier.

Post, E.J. [1962] *Formal Structure of Electromagnetics*, Amsterdam: North-Holland.

Ragusa, S. [1988] New first-order conservation laws for the electromagnetic field, *Nuovo Cimento B* **101**, 121.

Ramachandran, G.N. and S. Ramaseshan [1961] *Encyclopedia of Physics* **XXV/1**, Berlin: Springer.

Rao, B.R. and T.T. Wu [1965] On the applicability of image theory in anisotropic media, *IEEE Trans. Antennas Propagat.* **13**, 814.

Rumsey, V.H. [1954] Reaction concept in electromagnetic theory, *Phys. Rev.* **94**, 1483.

Rumsey, V.H. [1961] A new way of solving Maxwell's equations, *IRE Trans. Antennas Propagat.* **9**, 461.

Rumsey, V.H. [1964] Propagation in generalised gyrotropic media, *IEEE Trans. Antennas Propagat.* **12**, 83.

Satten, R.A. [1958] Time-reversal symmetry and electromagnetic polarization fields, *J. Chem. Phys.* **28**, 742.

Setlow, R.B. and E.C. Pollard [1964] *Molecular Biophysics*, Reading,MA: Addison-Wesley.

Shaw, T.I. [1972] The circular dichroism and optical rotatory dispersion of visual pigments, in *Handbook of Sensory Physiology, Vol. 7, Part 1* [Ed.: H.J.A. Dartnall], Berlin: Springer.

Shichi, H. [1971] Circular dichroism of bovine rhodopsin, *Photochem. Photobiol.* **13**, 499.

Shichi, H. [1983] *Biochemistry of Vision*, New York: Academic.

Shore, R. and G. Meltz [1962] Anisotropic plasma-covered magnetic line source, *IRE Trans. Antennas Propagat.* **10,** 78.

Silverman, M.P. [1985] Specular light scattering from a chiral medium: unambiguous test of gyrotropic constitutive equations, *Lett. Nuovo Cimento* **43,** 378.

Silverman, M.P. [1986] Reflection and refraction at the surface of a chiral medium: comparison of gyrotropic constitutive relations invariant or non-invariant under a duality transformation, *J. Opt. Soc. Am. A* **3,** 831; *ibid.* **4,** 1145; *ibid.* **5,** 1852.

Silverman, M.P. and R.B. Sohn [1986] Effects of circular birefringence on light propagation and reflection, *Am. J. Phys.* **54,** 69.

Silverman, M.P. and T.C. Black [1987] Experimental method to determine chiral asymmetry in specular light scattering from a naturally optically active medium, *Phys. Lett. A* **126,** 171.

Silverman, M.P., N. Ritchie, G.M. Cushman and B. Fisher [1988] Experimental configurations using optical phase modulation to measure chiral asymmetries in light specularly reflected from a naturally gyrotropic medium, *J. Opt. Soc. Am. A* **5,** 1852.

Smith, A.C. [1967] Waves in micropolar elastic solids, *Int. J. Engng. Sci.* **5,** 741.

Tellegen, B.D.H. [1948] The gyrator: A new electric network element, *Phillips Res. Repts.* **3,** 81.

Tinoco, Jr., I. and M.P. Freeman [1957] The optical activity of oriented copper helices. I. Experimental, *J. Phys. Chem.* **61,** 1196.

Tinoco, Jr., I. and R.W. Woody [1960] Optical rotation of oriented helices. II. Calculation of the rotatory dispersion of the alpha helix, *J. Chem. Phys.* **32,** 461.

Trelstad, R.L. [1982] Multistep assembly of Type I collagen fibrils, *Cell* **28,** 197.

Unz, H. [1964] Electromagnetic radiation in drifting Tellegen anisotropic medium, *IEEE Trans. Antennas Propagat.* **12,** 83.

Urry, D.W. and J. Krivacic [1970] Differential scatter of left and right circularly polarized light by optically active particulate systems, *Proc. Nat. Acad. Sci. USA* **65,** 845.

Vacatello, M. and P.J. Flory [1984] Helical conformations of isotactic poly (methyl methacrylate). Energies computed with bond angle relaxation, *Polym. Commun.* **25**, 258.

van Bladel, J. [1984] *Relativity and Engineering*, Berlin: Springer.

van Bladel, J. [1985] *Electromagnetic Fields*, New York: Hemisphere.

van Blokland, G.J. and S.C. Verhelst [1987] Corneal polarization in the living human eye explained with a biaxial model, *J. Opt. Soc. Am. A* **4**, 82.

van de Hulst, H.C. [1981] Light Scattering by Small Particles, New York: Dover.

van't Hoff, J.H. [1874] Sur les formules de structure dans l'espace, *Archives néerlandaises des sciences exactes et naturelles* **9**, 445.

Varadan, V.K. and V.V. Varadan (Eds.) [1980] *Acoustic, Electromagnetic and Elastic Scattering -- Focus on the T-Matrix Approach*, New York: Pergamon.

Varadan, V.K., V.V. Varadan and A. Lakhtakia [1987a] On the possibility of designing anti-reflection coatings using chiral composites, *J. Wave-Mater. Interact.* **2**, 71.

Varadan, V.K., A. Lakhtakia and V.V. Varadan [1987b] A comment on the solution of the equation $\nabla \times \mathbf{a} = k\mathbf{a}$, *J. Phys. A* **20**, 2649.

Varadan, V.K., A. Lakhtakia and V.V. Varadan [1987c] Scattering by beaded helices: Anisotropy and chirality, *J. Wave-Mater. Interact.* **2**, 153.

Varadan, V.K., A. Lakhtakia and V.V. Varadan [1988] Propagation in a parallel-plate waveguide wholly filled with a chiral medium, *J. Wave-Mater. Interact.* **3**, 267.

Varadan, V.K., A. Lakhtakia and V.V. Varadan [1987d] Radiated potentials and fields in isotropic chiral media, *J. Phys. A* **20**, 4697.

Varadan, V.V., A. Lakhtakia and V.K. Varadan [1986] Geometry can be the basis for acoustic activity (*a la* optical activity) in composite media, *J. Wave-Mater. Interact.* **1**, 315.

Varadan, V.V., A. Lakhtakia and V.K. Varadan [1987] On the equivalence of sources and duality of fields in isotropic chiral media, *J. Phys. A* **20**, 6259.

Varadan, V.V., A. Lakhtakia and V.K. Varadan [1988] Equivalent dipole moments of helical arrangements of small, isotropic, point-polarizable scatterers: Application to chiral polymer design, *J. Appl. Phys.* **63**, 280.

Voigt, W. [1887] Theoretische studien über die elastizitätsverhältnisse der kristalle, *Abh. Ges. Wiss. Göttingen* **34** (1887).

Waterman, P.C. [1969] Scattering by dielectric obstacles, *Alta Freq.* **38**, 348.

Waterman, P.C. [1971] Symmetry, unitarity, and geometry in electromagnetic scattering, *Phys. Rev. D* **3**, 825.

Weiglhofer, W.S. [1987] Scalarisation of Maxwell's equations in general inhomogeneous bianisotropic media, *Proc. IEE-H* **134**, 357.

Weiglhofer, W.S. [1988] Isotropic chiral media and scalar Hertz potentials, *J. Phys. A.* **21**, 2249.

Weiglhofer, W.S. [1989] A simple and straightforward derivation of the dyadic Green's function of an isotropic chiral medium, *Arch. Elektr. Über.* **43**, 51.

Winkler, M.H. [1956] An experimental investigation of some models for optical activity, *J. Phys. Chem.* **60**, 1665.

Lecture Notes in Mathematics

Lecture Notes in Physics

G.Scharf, University of Zurich, Switzerland

Finite Quantum Electrodynamics

1989. 240 pp. 4 figs. (Texts and Monographs in Physics).
ISBN 3-540-51058-3

Contents: Preliminaries. – Relativistic Quantum Mechanics. – Field
Quantization. – Causal Perturbation Theory.

G.Venkataraman, Ganesan, Hyderabad; D.Sahoo, Indira Gandhi Centre
for Atomic Research, Kalpakkam, Tamil Nadu; V.Balakrishnan, Indian
Institute of Technology, Madras, Tamil Nadu, India

Beyond the Crystalline State

An Emerging Perspective

1989. X, 207 pp. 87 figs. (Springer Series in Solid-State Sciences, Volume
84). ISBN 3-540-19110-0

Contents: Introduction. – Variety in Structures. – Order Out of Disorder.
– Defects and Topology. – Structures by Projection. – Beyond Simple
Geometry. – Tilings in One Dimension. – Ergodicity Breaking. – Symme-
try Breaking – A Second Look. – Appendix A–F. – References. – Author
Index. – Subject Index.

G.Vertogen, Catholic University of Nijmegen; W.H. de Jeu, Open
University of Heerlen, The Netherlands

Thermothropic Liquid Crystals, Fundamentals

1988. XI, 324 pp. 93 figs. (Springer Series in Chemical Physics, Volume
45). ISBN 3-540-17946-1

Contents: Mesomorphic Behaviour. – Continuum Theory. – Orientational
Order and Anisotropic Properties. – Liquid Crystalline Phases and Phase
Transitions. – References. – Index of Compounds. – Subject Index.

M.L.Cohen, University of California, Berkeley, CA; J.R.Chelikowsky,
University of Minnesota, Minneapolis, MN, USA

Electronic Structure and Optical Properties of Semiconductors

1988. XII, 264 pp. 161 figs. (Springer Series in Solid-State Sciences,
Volume 75). ISBN 3-540-18818-5

Contents: Introduction. – Theoretical Concepts and Methods. – Pseudo-
potentials. – Response Functions and Density of States. – Low Energy
Probes of Semiconductors. – Optical and Electronic Spectra of Semiconductors. – High Energy Probes of Semiconductors: X-Rays. – Diamond
and Zinc-Blende Structure Semiconductors. – Wurtzite Structure Semiconductors. – Chalcopyrite Structure Semiconductors. – IV-VI Semiconductors. – Triatomic, Layer, Chain, and Amorphous Models. – References. – Bibliography: Electronic Structure and Optical Properties of
Semiconductors. – Subject Index.

Springer-Verlag
Berlin Heidelberg New York
London Paris Tokyo Hong Kong